C. M. H. Barritt

Dip. Arch.

Advanced building construction

Volume 2

Second edition

Illustrated by the author

Longman
Scientific &
Technical

Longman Scientific & Technical,
Longman Group UK Limited,
Longman House, Burnt Mill, Harlow,
Essex CM20 2JE, England
and Associated Companies throughout the world

First published 1985
Second edition 1988
Fourth impression 1993

British Library Cataloguing in Publication Data
Barritt, C. M. H.
  Advanced building construction. — 2nd ed.
  Vol. 2
  1. Building
  I. Title
  690   TH145

ISBN 0-582-01970-2

Set in 10/11 Linotron 202 Times
Produced by Longman Singapore Publishers (Pte) Ltd.
Printed in Singapore.

Contents

Preface

This book is a companion to the second edition of *Advanced Building Construction*, Vol. 1. The two, taken together, provide all the information required by students taking building construction in the BTEC Higher National awards of Higher Certificates or Higher Diploma. As a subject, building construction is not confined to BTEC courses and these books should be of help to anyone taking other courses of study in which a knowledge of the techniques employed in building is required.

Each chapter deals with a different topic and in order to try to cover in a few pages what other authors have written a whole book about, I have concentrated on principles rather than particulars, uses rather than varieties and a review of the bases for selection rather than the complete range of choices available. More information can be obtained by reference to the books listed in the bibliography. As in the companion volume to this book, the illustrations are intended to indicate a method rather than to appear as a working drawing which might be taken to exclude other forms of construction. Therefore they are all freehand sketches and should be studied for their content rather than for their detail.

My son Paul has helped in this work by checking my manuscript for errors and I am grateful to him for again giving me this assistance. I am also indebted to Gill Barritt for typing the manuscript, which meant she had to decipher my handwriting, a task which she managed with commendable efficiency, and to my wife for all her help and encouragement without which this work might never have been finished.

My thanks too, to Colin Bassett for his welcome advice and for ensuring the accuracy of the technical content, to my colleagues at the Southend College of Technology for the assistance I received in assembling the information needed and, finally, to the staff at the Longman Group for all the care they take in producing the books in this series.

Acknowledgements

We are grateful to the Controller of Her Majesty's Stationery Office for permission to reproduce page 1 of HSE form F91 (Part 1).

Part I

General

Chapter 1

Building process legislation

1.1 The Construction Regulations

The manner in which work is carried out on site is governed by the standards laid down in four sets of Regulations:

The Construction (General Provisions) Regulations 1961
The Construction (Lifting Operations) Regulations 1961
The Construction (Working Places) Regulations 1966
The Construction (Health and Welfare) Regulations 1966.

The last of these, the Health and Welfare Regulations, was completely amended in 1974 and was further amended in 1980 by the Control of Food at Work Regulations and in 1981 by the Health and Safety (First Aid) Regulations. The other three apply as originally drafted.

In additi⌐⌐ *t*⌐ these main Regulations, several other Regulations have been a⌐⌐⌐oved covering specific topics, such as the use of woodworking machines, the protection of eyes, the use of abrasive wheels, the working of asbestos, etc.

All the Construction Regulations apply to any type of building work and most types of civil engineering and construction work. This generally includes site clearance, demolitions, foundations, the structure itself and all ancillary or finishing work.

Application can be made to the Factory Inspectorate for an exemption under these Regulations in respect of plant (e.g. a special type of hoist) or a class of work (e.g. steeplejacking) but it will be granted only where it can be shown either that compliance with the

Regulations is not necessary to maintain the safety of an employee or that it is not practicable.

1.2 The Construction (General Provisions) Regulations 1961

This Regulation deals with working conditions in connection with excavations, use of explosives, work adjacent to water, surface transport, demolitions and a number of miscellaneous guides for safe working practice. In common with all the other Regulations, due allowance must be made at the pre-tender stage for the cost of observing the requirements, and during the construction stage for complying with them.

1.2.1 Safety supervisor

A primary consideration arising from this Regulation is the need to appoint a safety supervisor when the contractor regularly employs 20 or more men in total (not just on one site). This means that any expansion of a builder's labour force to suit the work being undertaken has to be monitored carefully to see whether a safety supervisor is required and the cost then covered in the tender.

When appointed, the safety supervisor's duties are twofold: firstly to keep the management and staff advised on legislation applicable to the work in hand, and secondly to supervise the safe conduct of the work. His name must be shown on the abstract of the Factories Act or on the copy of the Construction Regulations which must be displayed on site or in the builder's office. He can cover the work on several sites and can, as is quite often done, be appointed to act for more than one builder.

1.2.2 Excavations

Almost without exception, all building and civil engineering work involves digging into the surface of the earth or, in some cases, tunnelling below it. What is encountered when this is done can vary immensely but, whatever is found, the work must be carried out in such a way that no one is put at risk.

The Regulations require that any work over 1.2 m deep where there is a risk of material collapsing or falling, must be adequately supported by proper timbering, trench sheeting, etc. This can be quite an expensive operation and due allowance must be made for the cost and the time involved.

Once work has started, the timbering must be properly constructed under competent supervision and maintained in good order and, where the depth exceeds 2 m, the excavation must be guarded. To ensure maintenance of safety, the timbering must be inspected every day by a competent person and thoroughly examined every seven days. A record of the thorough examinations must be kept (see Fig. 1.1).

Plant Equipment or Work	TESTING		EXAMINATION		INSPECTION	
	When it's to be done	Record on Form no.	When it's to be done	Record on Form no.	When it's to be done	Record on Form no.
Scaffolding					Weekly & after bad weather	91 (Pt.1) Sect. A
Excavations			Weekly	91 (Pt.1) Sect. B	Every day	
Coffer dams & Caissons			Weekly	91 (Pt.1) Sect. B	Every day & before working in them	
Cranes: Generally	Every 4 years	96	Every 14 months	91 (Pt.2) Sect. G	Weekly	91 (Pt.1) Sect. C
Cranes: Anchorage & ballast	After erection or any change	91 (Pt.1) Sect. D	After bad weather			
Cranes: Safe load Indicators	After erection	91 (Pt.1) Sect. E			Weekly	91 (Pt.1) Sect. C
Pulley blocks & gin wheels	Before use and after alteration	80	Every 14 months	91 (Pt.2) Sect. G	Weekly	91 (Pt.1) Sect. C.
Hoists (Passenger and goods)	Before use and after alteration or increase in travel height	75 91(Pt.1) Sect. F after alteration	Every 6 months	91 (Pt.2)	Weekly	91 (Pt.1) Sect. C
Chains, ropes, slings	Before first use (not for fibre ropes)	97 87 for wire rope	Every 6 months	91 (Pt.2) Sect. J		

NOTES:
Tests, examinations and inspections must be by a 'competent person'
Generally: Tests by manufaturer, erector or insurance company engineer
Examinations by ins. co. engineer or contractor's own engineer
Inspections by operator of equipment or other site staff
User must ensure that equipment has been checked by a competent person

Fig. 1.1 Plant tests, examinations and inspections

1.2.3 Use of explosives

The skilful use of explosives in the right circumstances can greatly shorten the life of a builder's contract. In the wrong hands it can shorten the life of a builder's employees. The Regulations require that the work must be carried out by shot firers fully trained in the use of explosives and experienced in the type of work to be done. The second point is as important as the first. For instance, a knowledge of explosives gained in a quarry or in the armed services would not qualify somebody to carry out, say, demolition work, with safety and efficiency.

Shot firers must be not less than 21 years of age and appointed by the site agent in writing. Explosives may be handled by people under the age of 21 but only under the supervision of a properly appointed shot firer, who alone is allowed to make the detonator connections.

To ensure the safety of site staff, a set of rules must be drawn up by the site agent and shot firer, allowing for warning notices and signals and the posting of sentries.

Further precautions must be observed and allowed for in the storage of explosives. Small quantities (up to 4.5 kg of blasting explosive or 13.6 kg of gunpowder) may be kept in a clearly labelled, stout wooden box fitted with a hinged lid and a lock. Detonators (up to 100) may also be stored in a separate, similar box. Over these quantities a completely detached substantial building is required, constructed and sited as specified by the Regulations.

The regular storage of explosives requires a licence to be obtained from the local authority and the use, even once, of explosives requires a certificate from the police. The storage licence must be displayed and the store clearly marked and under the charge of a responsible person who must keep a record of issues. The store, stocks and records may be inspected at any time by HM Inspector of Explosives, the local authority's inspector of explosives, a police inspector or an inspector of mines and quarries.

1.2.4 Work over, on or near water

The concern of the Regulations, and the great complex of other legislation on this subject, is twofold: firstly to try to minimise the risks and secondly to ensure that, if the worst does happen, rescue operations are swift and efficient.

Most safety precautions are similar to those to be observed on a normal building site – safe and secure working places and means of access to them (slippery surfaces must be guarded against particularly); proper illumination in dark areas or at night; and correct clothing such as non-slip footwear, safety helmets and buoyancy aids. The choice of buoyancy aids will be a compromise between freedom of movement and efficiency of flotation. The

Board of Trade life jacket, for instance, is very efficient but is intended to be worn in emergencies when abandoning ship and is too restrictive to work in. A wide choice is available, but it should be made in accordance with BS 3595: 1981.

The rescue equipment recommended comprises life buoys with lifelines attached, set at intervals along the workings (these should be a Board of Trade approved type); grab lines attached to the working place with marker floats at the free end and rescue boats which should patrol continuously while work is in progress.

Safety procedures should be formulated, such as a set routine for raising the alarm or for rescue drill, and men should work in pairs so that one can raise the alarm. The number of men at work should be periodically checked, and each man trained to be his own early warning system.

1.2.5 Surface transport

This part of the Regulations is largely concerned with the provision to be made for the proper construction, adequate safety and correct use of railway tracks, locomotives and rolling stock. It also covers the use of mechanically propelled vehicles and trailers to the extent that they are required to be in good working order and used correctly, that men are not allowed to ride on them in insecure positions, and that adequate provision is made to prevent a vehicle running over the edge of an excavation or embankment.

1.2.6 Demolitions

No great skill is required to demolish a building, but to do so in a safe, orderly manner calls for the training and experience of an expert.

In many cases, the value of the recovered material will cover the cost of the demolition work, but this consideration should not be allowed to affect the thoroughness with which the safety measures are applied.

Before the work starts a thorough examination should be made of the site (looking for wells, buried tanks, inspecting the type of ground, etc.), and also of the building to establish its form of construction and method of demolition (prestressed concrete members, for instance, require very different treatment to normal reinforced concrete work). It is also a required preliminary that the owner should inform the local authority and, if the work will exceed six weeks, HM Factory Inspectorate. The contractor is well advised to check that this has been done.

Proper hoardings, safety barriers and fences must be erected and the workmen issued with eye protection, safety helmets, dust masks, gloves, safety footwear, hearing protectors and, where there is a danger of falling, safety belts and harnesses.

There are six methods commonly used in demolition work: demolition by hand, by a swinging steel ball on a crane, by a pusher arm mounted on a suitable vehicle, by deliberate collapse where key structural members are removed, by pulling down with wire ropes, and by explosives. Whichever method is used, any parts of the structure left standing at the end of the day must be checked for stability. (*See Ch. 22 Advanced Building Construction Vol. 1.*)

1.2.7 Miscellaneous items

This part of the Regulations deals with a number of small but none-the-less important matters of safe working. In particular it covers the provision of guards to all moving and potentially dangerous parts of machinery; the safety of electrical supply cables (especially overhead wires where cranes, excavators, etc. are operating); protection from falling material; lighting of working places; the removal of danger due to material left with projecting nails (one of the most common causes of minor accidents) and loose materials obstructing platforms, gangways, etc., the avoidance of lifting excessively heavy weights; and a number of others more fully dealt with elsewhere.

1.3 The Construction (Lifting Operations) Regulations 1961

These Regulations are concerned with the safe construction, installation and use of all lifting appliances from a humble gin wheel lashed to a scaffold, to a huge tower crane, hoists for passengers and materials, and the carrying out and recording of inspections.

The employment of many of the pieces of equipment covered by these Regulations is a subject of careful consideration in the pre-tender stage from both the cost-effective aspect and the site planning requirements. Due regard must be paid to the implications of the safety standards laid down and allowance made for the cost of inspecting and keeping the records stipulated.

In the pre-construction period the site layout must be arranged round the principal piece of lifting equipment, allowing space for the clearances specified in the Regulations.

Once work has started on site then the equipment must be organised, correctly erected, properly used and regularly inspected. The Regulations place the responsibility for compliance with the user rather than the owner (in the case of hired equipment), and also charge him with weekly inspections, anchorage and stability tests, automatic indicator tests and passenger hoist tests, the records of which must be kept on site for the Factory Inspector to see.

1.4 The Construction (Working Places) Regulations 1966

It is under the rules laid down in these Regulations that the form, construction and safety of scaffolding and the means of access thereto is determined. Due allowance must be made for both the cost of providing the correct scaffold for the work in hand and for the space in which to erect it.

An interpretation of the Regulation that 'every work place must be safe' is that it must be large enough for a man to stand, move about, carry out his work and, if necessary, stock the materials he is using. The scaffold must be properly tied to the building and fitted with guard rails and toe boards.

Every employer is required to see that the Regulations which apply to his men are observed, and where more than one employer is using the same scaffold each is responsible for his own men. The Regulations also charge every employee with the duty to obey the requirements and to report any defects to his foreman or his safety officer,

A safe route must be provided to the scaffold. Decisions must therefore be made, when estimating the scaffold costs to be allowed in a tender figure, as to whether a simple ladder access will be adequate or a properly constructed scaffolding staircase provided. The latter is usually put in when the site accommodation buildings have to be located on the scaffold.

A ladder is a piece of builder's equipment which is frequently used and almost as frequently abused. It must always be set at the correct angle of about 75° (1 out for every 4 up), it must be lashed securely at the top and it must rise at least 1.1 m above the stepping-off point at working level.

Scaffolding must be erected under the supervision of a trained person. It must be inspected before it is used and only a trained person should alter or dismantle it. A similarly trained and experienced person must also inspect the scaffold once every week and in addition after any rough or cold weather which might have affected its stability. This is important, as frost can separate joints and slacken couplers, changes in temperature can affect the grip of reveal ties, and high winds can both rattle the scaffolding loose and ease it away from the building.

Each of these reports must be recorded in a scaffold register form F91 Part 1 Section A, (see Fig. 1.2). If the job is not expected to last more than six weeks this register can be kept in the builder's office. Otherwise it must be kept on site and available for the factory inspector to see.

Factories Act 1961

Construction (Working Places) Regulations 1966

SCAFFOLD INSPECTIONS

SECTION A

Name or title of employer or contractor ....

Address of site ....

Work commenced—Date ....

Reports of results of inspections under Regulation 22 of scaffolds, including boatswain's chairs, cages, skips and similar plant or equipment (and plant or equipment used for the purposes thereof)

Location and description of scaffold, etc. and other plant or equipment inspected (1)	Date of inspection (2)	Result of inspection State whether in good order (3)	Signature (or, in case where signature is not legally required, name) of person who made the inspection (4)

NOTES TO SECTION A

(1) *Short check list—at each inspection check that your scaffolding does not have these faults:*

		Week 1 2 3 4
FOOTINGS	Soft and uneven	
	No base plates	
	No sole boards	
	Undermined	
STANDARDS	Not plumb	
	Jointed at same height	
	Wrong spacing	
	Damaged	
LEDGERS	Not level	
	Joint in same bays	
	Loose	
	Damaged	

		Week 1 2 3 4
BRACING	Some missing	
'Facade and ledger'	Loose	
	Wrong fittings	
PUTLOGS and TRANSOMS	Wrongly spaced	
	Loose	
	Wrongly supported	
COUPLINGS	Wrong fitting	
	Loose	
	Damaged	
	No check couplers	
BRIDLES	Wrong spacing	
	Wrong couplings	
	Weak support	

		Week 1 2 3 4
TIES	Some missing	
	Loose	
BOARDING	Bad boards	
	Trap boards	
	Incomplete	
	Insufficient supports	
GUARD RAILS & TOE BOARDS	Wrong height	
	Loose	
	Some missing	
LADDERS	Damaged	
	Insufficient length	
	Not tied	

(2) *This check list is not part of the report required by Regulation 22: see also para 5 of Notes on cover page (ii) and page 13.*

Fig. 1.2 Scaffold Register from Form F91

1.5 The Construction (Health and Welfare) Regulations 1966

This set of Regulations is concerned with the provision of on-site shelter, for protection and meals, washing and sanitary facilities. The provisions of first-aid equipment was formerly covered as well but this was repealed and replaced by the Health and Safety (First Aid) Regulations 1981, which came into effect on 1 July 1982.

Messroom and sanitary accommodation is related in its size to the anticipated total number of people employed on the site at any one time. Separate sanitary facilities must be provided for men and women but employees of different firms can share the same facilities. It is necessary, therefore, to estimate the maximum number of on-site staff in the contractor's employment as well as those working for subcontractors, and to make due allowance both for the cost and the space required.

Where facilities are shared, the employer providing them (usually the main contractor) must keep a register (Form 2202) showing what is being shared and the names of the firms sharing them; and he must give the associated certificate (Form 2202 Part A) to each of the other employers showing what is being shared. This register must, like the others mentioned previously, be available for examination by the Factory Inspector.

Chapter 2

Site mechanisation

2.1 Construction plant

The operations on a building site are mainly those of demolition, excavation, transporting materials, preparing materials, hoisting materials and men, and fixing or assembling. All these can be done by hand and equally they can all be done by, or with the assistance of, mechanical plant. The contractor's problem is to decide which will be the most economic.

At the extremes, the decision is not difficult – a small domestic extension will be predominantly a hand operation, whereas a multi-storey office block will rely heavily on mechanical equipment. It is with the medium-sized contract where either method could be the best, and the wrong choice could consume the builder's profit.

2.2 Selection of plant

Many factors influence a decision to use, and the selection of, machinery on a building site. The most important are the nature of the work (size and type of operations), the construction method proposed and the time allowed.

With respect to the first, even quite small jobs can take advantage of the benefits of mechanisation. Very few builders mix concrete or mortar by hand, and even a small house may contain 100–150 tonnes of material which may be handled more than once, making some form of transport and hoisting a feasible proposition. Where the overall size of the contract clearly calls for mechanical aids, the size of the individual operations – how much excavation, how much concrete, how high the lifting – must be studied in order to match the equipment to the work and maintain an optimum output. The type of operation, for instance new work or restoration, will determine the programme to be followed and from this can be worked out a schedule for the plant and equipment required.

The method of construction – prefabricated or on-site work, large or small building units – will indicate the type and capacity of equipment to be used. In many cases, where the system proposed cannot be carried out without the use of machinery, the method of construction will be matched to the plant available.

The contract period and finishing date are most important considerations, either from the point of view of cost to the contractor and to the building owner or because the building must be available for use on a certain date. The right amount of mechanisation can speed up the process of building as well as saving costs; more equipment can speed it up further, but usually the cost will also go up. Where a tight programme must be followed, this increased expenditure may be acceptable.

2.3 Site layout

The storage of materials and their movement, both horizontally and vertically, usually determine the layout of the site around the building during the construction period, with the aim of reducing the amount of handling.

Storage areas are located as near to the point of delivery as practicable; processing areas (concrete mixing, reinforcement fabrication, on-site joinery shop, etc.) placed conveniently close to the storage areas; and means of transportation of the delivered or processed materials arranged to take the shortest route.

The type and quantity of each material to be stored, the work to be carried out on it, if any, and the method of transporting it must all be carefully studied, since the dictates of each will affect the final layout plan.

2.4 Earth-moving and excavation plant

The excavation of the ground and the disposal of the products of that excavation are activities well suited to mechanisation because the operation is largely the same on all sites, only varying in extent. Furthermore, the work involves a weight of material which powered equipment can handle more efficiently and economically than can be achieved by hand working. Only in awkward and constricted working conditions or where site access is severely restricted in width is hand excavation considered, and even to deal with this small machines have been developed which can be negotiated through a domestic doorway.

The equipment used for builder's excavations (civil engineering works employ additional and larger machines) consists of a power unit, usually diesel engined, mounted on wheels or caterpillar tracks with a variety of attachments operated either by ropes or hydraulically.

In operation, the excavator must either dig below its wheels or tracks, in which case a backacter is used; or it must dig the ground in front and above its wheels, when a face shovel is used. Alternatively it may be required to remove the surface of the ground down to a formation level for which purpose a scraper, skimmer or loading shovel would be employed (see Fig. 2.1).

In detail and capacity, the range of equipment available is large but, because of its versatility, the multi-purpose excavator is very popular with small to medium-sized building firms. It comprises a power unit, generally diesel engined, mounted on wheels and fitted with a back acting bucket at the rear of the machine and a loading shovel at the front. Both the bucket and the shovel are hydraulically operated. A tracked version is available but this cannot be driven on the highway and therefore the cost of transportation to site is increased. Back acting buckets (sometimes called hoes) are available in a variety of widths to suit the trench being excavated, giving a range of capacities from 0.1 to 1.0 m^3. Large machines can excavate to a depth of 6 m and have a forward reach of 7.5 m.

A face shovel (also called a luffing or crowd shovel) can work on a face up to 10 m high and has a bucket capacity from 0.3 to 3 m^3 which can be slewed through 360° and is usually emptied by opening a door in the bottom.

Scrapers possess the largest capacity for reduced level working, up to 20 m^3. Skimmers have a bucket which contains 0.5 m^3 but they can be very accurate in operation and loading shovels can haul up to 2 m^3. Loading shovel capacities of up to 5 m^3 can be used when the equipment is employed for the purpose its name implies, that of shovelling up loose material and loading it into transport.

Transportation of the excavated material is achieved with dumpers, lorries or skips, the choice of which is based on the quantity to be moved, the rate of production and whether it is on or off site. Dumpers are ideal for small quantity (about 0.5 m^2) short haul transport on site, and skips are particularly economical when the rate of production is slow (hire is charged per day or week and the maximum skip capacity is 4.5 m^3). The most common form of transport of excavated material is the lorry, but care must be taken

14

in loading as damage to the suspension can be incurred by the shock of large quantities of spoil being tipped into the lorry body.

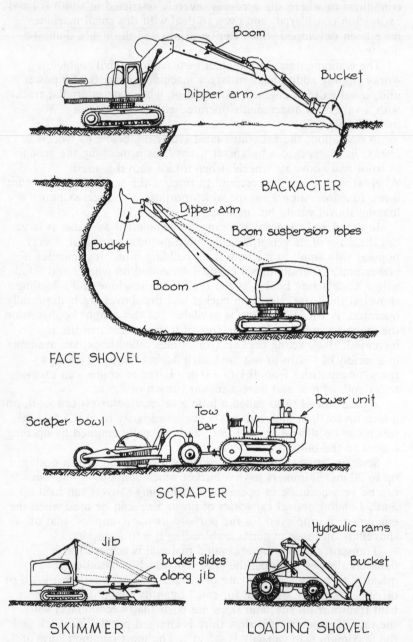

Fig. 2.1 Excavators

2.5 Materials and component handling plant

Plant provided for the purpose of moving materials and components generally either moves it horizontally or vertically (see Fig. 2.2). Exceptions to this are concrete pumps, dealt with in the next section, and fork-lift trucks, both of which transport materials in both directions.

At the lowest end of surface transport is the humble but ubiquitous barrow, followed by the power barrow. This latter is really a small version of the dumper, carrying up to 0.4 m^3 and steered by the rear single wheel of the three-wheeled chassis, under

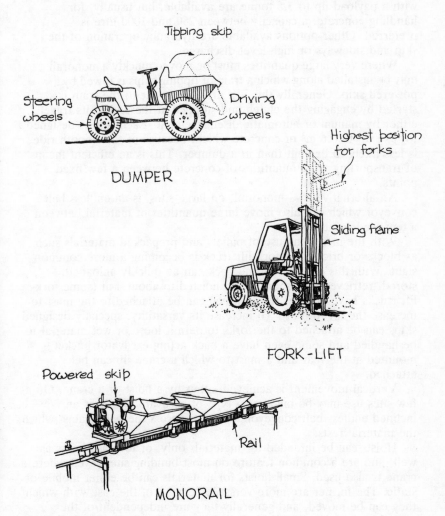

Fig. 2.2 Surface transporters

the control of a pedestrian. The ground needs to be fairly firm and not excessively steep for these power barrows to work and they suffer the disadvantage of being too large to fit into most hoists.

The plant most frequently used for surface transport on site is the dumper. Dumpers are called upon to handle a wide variety of materials from wet concrete to drainpipes and excavated spoil to sanitaryware. They are usually diesel engined with a choice of two- or four-wheeled drive and with a skip mounted at the front which discharges forwards by gravity when a catch is released. Because of the wide variety of loads to be carried, the capacity of the skip is quoted as payload, water level, truck level and heaped. Dumpers with a payload up to 3.5 tonne are available, but usually, for handling concrete, a capacity between 250 and 1000 litre is preferred. Other options available are hydraulic operation of the skip and sideways or high-level discharge.

Where very large quantities must be moved quickly a monorail may be installed along which a train of tipping skips is towed by a powered skip. Generally this is a driverless system: the train is started by engaging the clutch and stopped at the end of its travel either by manual or automatic disengagement. Each skip is designed to take up to 0.6 m³ of concrete which, because of the smooth ride, is less likely to be spilt than in a dumper. This is an efficient means of transporting large quantities of concrete between a few fixed points.

An alternative to a monorail, on large sites, is an endless belt conveyor which can also move large quantities of material between a few fixed points.

With the increasing use of pallets and prepacked materials such as blocks or bricks, the fork-lift truck is becoming a more common sight. With this equipment, the pack can be quickly unloaded, stored, retrieved, transported and hoisted to about 4 m (some fork-lift trucks have another frame which can be attached to the mast to increase the lift to 6 m). To increase its versatility, specially designed skips can be attached to the forks to permit loose or wet material to be handled and some even have a back acting excavator bucket mounted at the rear and a mast to which a crane jib can be attached.

Vertical movement is achieved either by a hoist or a crane. On a few sites use may be made of a type of excavator which is an inclined endless belt conveyor with slats across the belt against which the material rests.

Hoists can be intended for materials only, or for passengers as well, and are a common feature on most building sites, even where a crane is also used. Small hoists for materials can be either mobile or static. The former are more versatile because of the ease with which they can be moved, and generally they are independent of the scaffold but are limited in their lifting capacity. Static hoists are

attached to the building or scaffold and can carry greater loads and
passengers to greater heights. All hoists must stand on a firm base,
be plumb and securely enclosed by protective wire mesh clipped to
a scaffold tube tower (see Fig. 2.3).

The development of larger building units and prefabricated
components and the economic pressures to build quickly have
combined to make a crane a much more common feature on a
building site than it used to be, even though it is an expensive piece
of equipment.

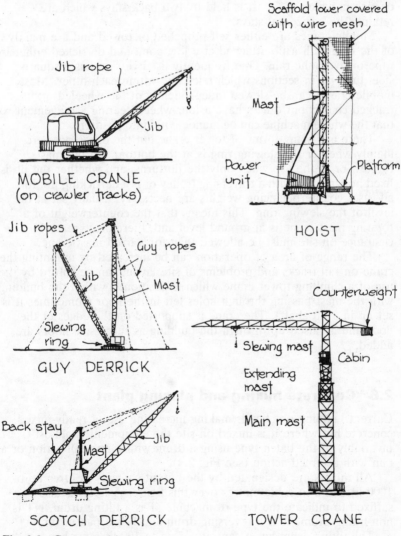

Fig. 2.3 Cranes and hoists

There are many ways in which cranes can be classified but the clearest division is derrick cranes, mobile cranes and tower cranes (see Fig. 2.3).

Derrick cranes are one of the older forms of crane and can handle their maximum load over a greater range of radius than many other forms of crane. They consist of a mast mounted on a turntable or slewing ring to the base of which is attached a latticed jib which is raised or lowered (derricked) by a rope from the top of the mast. There are two types of derrick crane, the guyed derrick in which the mast is supported by three or more guy ropes, and the Scotch derrick in which the mast is held by two back stays which are retained by horizontal stays.

Mobile cranes are either self-propelled or towed and are mainly of the type with a jib attached at a low point and derricked either by rope or hydraulic ram. One frequently used version has a square steel tube jib in sections which telescope within each other. Most mobile cranes can be slewed independently of the wheeled or tracked chassis but others have a four-wheel steering arrangement so that the whole machine can be manoeuvred into position.

There are many forms of tower crane but the differences are mainly whether the slewing ring is at the bottom and the mast rotates, or at the top when only the jib turns, and whether the jib is fixed horizontally with a traversing trolley or is a derricking jib. In all cases counterbalancing weights are necessary, usually set at the level of the slewing ring. This means that the counterweight of a rotating mast crane is at ground level and, therefore, greater clearance on site must be allowed than for a fixed mast crane.

The range of area of operation can be increased by mounting the crane on rail tracks and problems of site layout can be solved by the use of a climbing tower crane which is positioned within the building with the mast passing through holes left in the floor (sometimes it is set up in a lift shaft). The crane is supported on the edges of the floor and is raised up through the building as successive floors are added.

2.6 Concrete mixing and placing plant

Current building practice is making increasing use of ready-mixed concrete but where it is mixed on site the equipment is almost invariably of the batch type using a drum with a free-fall action or a pan with a forced action (see Fig. 2.4).

All mixers are designated by the output capacity in litres up to 1000 litre and in cubic metres over this figure. A letter is also suffixed to indicate the type of machine: T is a tilting drum, NT is a non-tilting drum, R is a reversing drum and P is a pan mixer.

The tilting drum has a capacity of up to 100 litre for hand feeding and up to 200 litre for power-operated loading shovels.

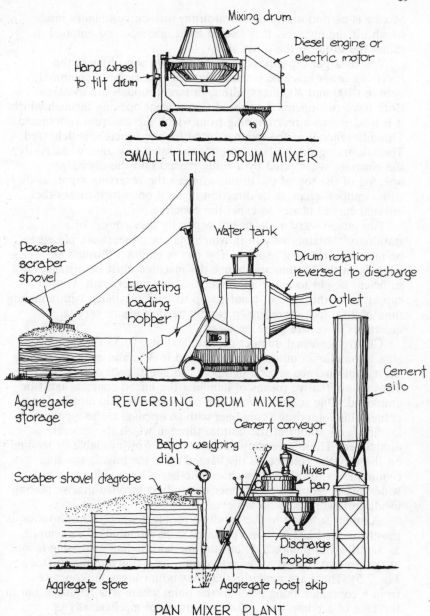

Fig. 2.4 Concrete mixers

Mixing is carried out in a conical drum with curved blades inside which lift the mixture. It is loaded from one side and emptied by tilting the drum to the other side.

Where larger capacities are called for, the non-tilting or the reversing drum machine is used. Both have a capacity commonly between 200 and 400 litres, although larger models are available. Both have two openings to the drum, a front opening through which it is loaded and a rear opening from which the concrete is obtained. The difference is in the manner by which the concrete is delivered. The non-tilting drum always rotates the same way and, when ready, the concrete is collected by a chute passed into the discharge opening of the top of the drum; whereas the reversing drum, as its name implies, changes its direction of rotation, which causes the internal curved blades to expel the concrete.

Pan mixers vary in detail but generally they consist of a stationary container or pan in which the constituents are placed to be mixed by rotating paddles. The pan is emptied through an opening in the bottom and hence the machine must be installed at a sufficient height to allow a dumper to drive underneath. Output capacities of this type of plant go up to 2 m³ per batch with a mixing time of 30 s and, therefore, it is only needed where very large quantities of concrete are required.

Clearly a normal dumper of, say, 500 litre (0.5 m³) will not be able to handle an output of 2 m³ mixed in 30 s and even over a short haul distance several dumpers can get in each others' way. Thus an alternative means of handling the mixed concrete must be employed. One solution is the use of a concrete skip attached to a crane. This comprises a container with an opening at the top for loading, and a door at the bottom through which the concrete is discharged. There are many variations and sizes available depending on the way the concrete is discharged from the mixer, and the requirements for placing, such as whether a large volume over a wide area is required or whether carefully controlled precise placing inside formwork is to be achieved.

Another solution to the handling problem where transportation is mostly on the level, is to use powered skips mounted on a monorail, as described in section 2.5, but the method finding increasing favour at the present time is a concrete pump, or a concrete placer (see Fig. 2.5). These are similar in that they both convey the concrete from a container along pipes to the point where it is required, but in the case of a pump this is done by means of mechanically or hydraulically operated pistons, whereas a placer impels the concrete by means of compressed air.

The supreme advantage of this method is the ease with which obstacles can be negotiated. For instance, ready-mixed concrete can be discharged from a lorry in the road into the hopper of a mobile pump and then fed through pipes over the site hoarding and across

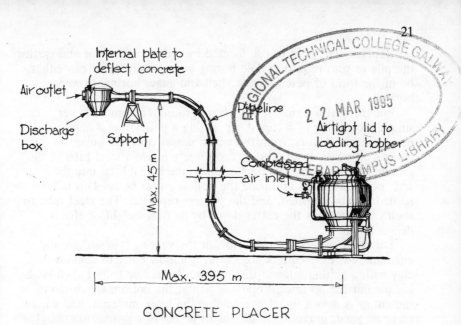

CONCRETE PLACER

Fig. 2.5 Concrete placer

the basement excavations to stanchion bases on the far side of the site. Horizontal delivery distances of as much as 450 m and 85 m vertically are attainable, but at these extremes the performance of the equipment, the choice of pipe size and the quality of the concrete must be very carefully studied.

This method achieves the highest rate of concrete delivery and, to make it economic, both the production or delivery of the concrete and the preparation for its placing must be matched to its capabilities.

2.7 Pile driving and installation plant

Since there are two distinct piling methods, displacement piles and replacement piles, two different sets of plant are available for their installation (see *Advanced building construction*, Vol. 1 Ch. 7). Displacement piles (those formed first and then inserted into the ground) are normally driven by a pile hammer which is raised and dropped onto the pile cap. The hammer travel is guided by a leader hung from a mobile crane or by a pile driving frame. The method, although simple, is noisy and creates vibrations in the ground, both of which disturb adjoining property owners and their buildings. Alternative methods are vibrating pile drivers which, by inducing vibrations in the pile temporarily reduce the frictional resistance, and the pile sinks into the ground under its own weight; or the hydraulic ram pile drivers, which act against an existing foundation or other installed piles and force the pile down into the ground.

Replacement piles (those formed by removing the soil and casting the pile *in situ*) require a hole boring arrangement. This can either be in the form of percussive or shell-and-auger boring, or auger boring.

For the first alternative the above-ground equipment is very simple, consisting of a tripod supporting a pulley and a hand or power winch. A rope from the winch passes over the pulley to the cutting bit which is allowed to drop freely inside a steel tube in the ground. As the cutter hits the bottom of the hole it bites into the soil, which is forced up inside the hollow cutter body. This is then hauled up and emptied, and the process repeated. The steel tube or shell either follows the cutter down by its own weight or else is driven down.

Different cutters are made to suit the varying types of ground encountered. Clay cutters are simply a hollow body to contain the clay with a cutting edge at the bottom. A sand or ballast shell is also hollow but has an inward opening flap at the bottom which closes as the cutter is drawn up, thus retaining the loose material, and where compact sand, gravel or chalk has to be bored a percussion chisel is used first to break up the soil, followed by a ballast shell to remove the debris.

Once the hole is bored and pouring of concrete starts, the above-ground equipment can be used to winch the tubes up out of the ground as the bore hole fills.

For certain cohesive soils and for short bored piles, an auger is used. This is like a large version of a bit from a carpenter's brace and bit, and acts in exactly the same way to bore a hole into the ground. For very small applications can be turned by hand, but usually it is mounted at the rear of a mobile mechanical auger machine.

2.8 Pumping and dewatering plant

There are three types of pump – displacement, centrifugal and submersible – each with their respective advantages and disadvantages, which must be appreciated before the correct choice can be made.

Displacement pumps consist of either a piston thrusting backwards and forwards in a cylinder or a flexible diaphragm sealed to a cylinder and moved up and down by an actuating lever. In both cases, a valve opens on the backward or upward stroke to admit water from the suction pipe, and on the forward or downward stroke the inlet valve closes and another opens to let the water into the delivery pipe. Since the water is only expelled during half the operating cycle, a pulsating delivery results. To achieve a more even flow (and to increase the capacity) two cylinders are used linked by a single arm so that each discharges alternately.

The diaphragm pump is a popular form with builders. Its capacity is often adequate for the volume of water to be pumped and it can raise it to a head of about 6 m. A great advantage of this pump over others is that it can handle water with up to 15 per cent solid matter in it – an important factor when, as is often the case, excavations need to be pumped dry.

Large volumes of clean water can be more efficiently moved by a centrifugal pump. This consists of an impeller with curved blades rotating at high speed within a close-fitting casing. The blades force the water outwards to the delivery pipe which is replaced by water drawn in at the centre from the suction pipe. This produces a continuous discharge but only works when the pump is full of water, and thus requires to be primed each time it is started. There is a self-priming version which retains a reservoir of water within the pump when the supply stops. It is a lightweight machine which is easy to maintain but it cannot handle water with solids in it.

In all pumps there is the lift side where the water is sucked in and the force side where the water is expelled. The lift is achieved by reducing the pressure at the end of the suction pipe, which causes the water to rise up the pipe by atmospheric pressure on the water surface. Thus the height of the lift (the suction head) which can be achieved is not only a question of the ability of the pump but also depends on the atmospheric pressure. The performance of the force side (the delivery head) is solely controlled by the efficiency and power of the pump. Where the water to be raised is at a low level, it is sometimes necessary to lower the pump to it to reduce the suction head; but usually the only surface other than the ground, on which the pump can stand, is at the bottom of the water. For this situation submersible pumps have been developed.

Submersible pumps are almost invariably electrically powered and generally of the centrifugal type, although a diaphragm pump version is available where solids are present in the water. They are capable of raising water up to a head of 30 m.

Besides being called upon to remove water collecting in excavations, pumps are also used to drain the ground before excavation starts. For this purpose a number of 50 mm tubes are sunk into the ground at about 2 to 3 m from the excavation at between 0.5 and 1.5 m centres (depending on the degree of sub-soil permeability). The bottom end of each pipe is connected to a perforated mesh-covered tube called a well point, and the tops are linked to a header pipe connected to a pump. The operation of the pump drains water into the well point, thus dewatering the sub-soil and allowing excavation to proceed in dry conditions, even, in certain circumstances, without the need for expensive timbering.

Chapter 3

Site and soil investigation

3.1 Definitions

Site investigation and soil investigation are two separate activities,
usually carried out at different times in the building programme (soil
investigation being the later) for different purposes and by different
people.

The examination of the site, to investigate the effect its features
will have on the design and construction of the building, is
frequently described as a site survey (although this may only be part
of the exercise) and is concerned with the topography of the site, its
surface features and the character and characteristics of its
surrounding areas. Soil investigation, as its name implies, probes
beneath the surface to ascertain the type, quality and extent of the
sub-soil.

3.2 Site survey

Before any detailed design work can be started, a great deal of
information is required concerning the site to be developed. This,
clearly, is an exercise occurring right at the beginning of the design
process, but before the builder can carry out any work planning,
either in the preparation of a tender or for the purpose of
programming the construction work, he will also need a site survey
which looks at the aspects of the site which affect the execution of
the building work rather than the design.

The shape of the surface is obtained from levels taken at points on the surface, usually on a grid pattern, by the use of a dumpy level, a theodolite, or as part of the readings taken with EDM equipment. Surface features are recorded and measured as the survey is conducted. The position, species, size and condition of trees is important, affecting, as it does, both the design and siting of the building and the type and depth of the foundation; the dimension and conditions of all man-made features such as buildings, roads, boundary walls or fences, etc., must be noted, as also must natural features such as rivers, ponds, rocky outcrops, etc.

This would all be recorded in a site survey drawing showing the shape of the site, details of the features and spot heights and contour lines. Where it is considered necessary, the surveyor may also include sections drawn through the site.

A site survey for a pre-tender report or construction programme would not be concerned with any of the subjects of the survey described above. For this the purpose and objectives are quite different, and in any case the builder will have the information already obtained, shown on the survey drawings. The function of this survey is to discover and assess all conditions and constraints which could affect the execution of the contract. This investigation would be concerned with the means of access, if any, to the site for heavy vehicles; any parking and loading restrictions; any foreseeable problems with movement of vehicles on the site; any tall buildings which may affect the use of a tower crane; availability of water, electricity and telephone connections; and it may even extend to enquiries into possible local sources of building materials and availability of labour. The information gleaned from this type of site inspection would be put down in a report, possibly using a standard form of the type shown in Fig. 3.1.

The objectives and extent of the site survey must be carefully matched to its purpose. If a preliminary study of the feasibility of developing the site is being made, then the amount of information needed is limited and it is usually necessary to keep costs to a minimum. If, however, the building is being worked out in its final and detailed form, every item of information which could affect the way the structure and foundations are designed is essential.

There are three main aspects of a site survey: the boundaries of the site, the shape of the surface and the features on that surface. When small, the dimensions of the site can be ascertained by direct measurement by tape or chain; when the site is large, the surveyor will resort to an optical survey using a theodolite, or to aerial photographs, or to the latest techniques of electronic distance measurement (EDM). This last system will calculate the distances directly in the field or the readings can be recorded on a tape which is subsequently fed into a computer linked to plotting equipment which draws the survey automatically.

Pre-Tender Site Visit Report

Project:

Tender no.:
Date:
Prepared by:

SITE INVESTIGATION
1. Availability of site:
   1.1 key holder   _____
2. Location:
   2.1 distance from head office   _____
   2.2 local transport services   _____
3. Access to site:
   3.1 road width   _____
   3.2 existing access   _____
   3.3 crossings   _____
   3.4 temporary roads   _____
4. Working space available for:
   4.1 offices, canteen, stores   _____
   4.2 materials, storage, handling   _____
   4.3 construction purposes   _____
   4.4 plant (and recovery)   _____
5. Services/location:
   5.1 toilets   _____
   5.2 water   _____
   5.3 electricity   _____
   5.4 telephone   _____
   5.5 any available from client   _____
6. Nature of ground:
   6.1 surface   _____
   6.2 bore holes   _____
   6.3 levels   _____
7. Site clearance:
   7.1 trees   difficulties   _____
       streams   recommended _____
       existing items methods   _____
   7.2 room for excavation plant   _____
   7.3 tips and charges   _____

LOCAL CONDITIONS
8. Availability of local labour:
   8.1 type and quality   _____
   8.2 inducement paid in area   _____
9. Labour Exchange:
   9.1 address   _____

   9.2 telephone number   _____
10. Existing damage in:
    10.1 adjacent buildings   _____

10.2 adjacent roads   _____

10.3 adjacent footpaths   _____

11. Local resources:
    11.1 builders merchants   _____
    11.2 sub-contractors   _____
    11.3 plant hire services   _____

12. Competition:
    12.1 other main contractors
         locally   _____

13. Adjacent properties:
    13.1 owners   name   _____
                      address   _____

                      name   _____
                      address   _____
    13.2 occupiers   name   _____
                      address   _____
                      name   _____
                      address   _____

SECURITY

14. General requirements:
    14.1 fencing, hoardings, etc.   _____
    14.2 fans   _____
    14.3 lighting   _____

15. Watchman/guard dogs
    15.1 playgrounds near   _____
    15.2 schools near   _____
    15.3 general local character   _____

OTHER CONDITIONS

16. Implications of:
    16.1 contract damages clause   _____
    16.2 phased working conditions   _____
    16.3 special insurances   _____

17. Special care required when
    pricing   _____

18. Possibility of flooding   _____
19. Special fire precautions   _____
20. Special site safety measures   _____
21. Work to be done by Client   _____
22. Other firms tendering   _____

22. Additional information   _____

Fig. 3.1 Pre-tender report form

3.3 Soil investigation

As with a site survey, any soil investigation exercise must have
regard to the purpose to which the information will be put and also
the amount of money which it is sensible to spend on this work.
Properly and fully carried out, a sub-soil investigation can be a very
expensive item – even a few trial holes can cost a surprisingly large
sum – and this, along with all the other expenses, will usually reduce
the amount of money available to be spent on the actual building. It
is, nevertheless, an important subject which should not be skimped.
The degree of sophistication in the methods used varies widely,
ranging from a few trial holes taken out by an excavator, to complex
on-site tests and highly scientific laboratory analysis of carefully
removed undisturbed soil samples and including electronic and
seismological methods of investigation.

On-site tests are designed either to assess the sub-soil as it exists
in its natural state, or to remove samples for examination from
which the performance of the sub-soil generally can be deduced.
Each has its limitations, the first by the difficulty of matching the
tests to the actual conditions to be created by the building and the
second by the possibility that the samples taken are not
representative of the majority of the sub-soil.

The most direct on-site test is one which applies a pressure to the
sub-soil to create the conditions the foundations will impose. This,
the plate load test, (see *Advanced building construction*, Vol. 1
Ch. 4) would seem to be a sensible way of finding out how the sub-
soil behaves under load, but caution and experience are required in
the interpretation of results because the test cannot reproduce either
the magnitude or the extent of the buiding load, and as a result the
test stress does not penetrate to the depth eventually to be loaded.
If there is a different stratum at this lower level with an inferior
bearing capacity, the ability of the sub-soil generally to carry
foundation loads will be less than the test predicts.

Another on-site test is the vane test (see *Advanced building
construction*, Vol. 1 Ch. 4). This is used to measure the shear
strength of soft cohesive soils and is applied by test equipment
consisting of a rod with four blades at the bottom end, at the top a
means of rotating the rod, and a torque gauge. The rod is forced
down into the ground and at predetermined depths is rotated at the
rate of one revolution per minute against a spring. Eventually the
soil shears and the blades rotate. The torque required to achieve this
is measured and recorded for interpretation by the foundations
engineer.

A compression test which can be carried out on site on extracted
samples of sub-soil has been developed by the Building Research
Establishment (Test 20, BS 1377). The sample is unconfined during
the test and usually fails by either bulging or shearing. Although this

is unrealistic, since neither can occur in the natural state, allowances can be made and usable information collected.

The most useful exercise on site is the examination of trial holes by an experienced surveyor. This will reveal the strata pattern, thickness and depths (down to the depth of the trial hole), soil types, water table level, and the quantity and description of water entering the hole.

To be of any value, the trial hole must be taken down to at least 1 m below the anticipated foundation level. If the loads of the building are large, their effect will carry further into the ground and the trial hole should be correspondingly deeper. The economic limit to depth is about 5 m and if the ground conditions, anticipated building load or proposed foundation type require an investigation to a deeper level, bore holes must be used.

General investigative borings should be taken down to a depth equal to the width of the building or 30 m, but where tests relate directly to foundations, it is not necessary to go deeper than 1.5 times the breadth of a pad foundation or 3 times the breadth of a strip foundation, since at these points the bulb of pressure has spread so that it is reduced to 20 per cent of the load at the foundation level (see Fig. 3.2).

The samples obtained from bore holes should be carefully sealed in an airtight container and sent to a testing laboratory. To each sample is attached a label giving details of the sample, the bore number, the position of the sample in the bore, and whether it is disturbed or undisturbed.

As well as obtaining samples, the process of boring can be interrupted to carry out standard penetration tests (Test 19, BS 1377). These serve as an additional guide to the relative density of the ground and hence its strength. The test consists simply of recording the number of blows from a 65 kg weight dropped 760 mm required to drive the tube down 300 mm (after having first driven it 150 mm).The way this test is interpreted is shown in Fig. 3.3. A modification to this is the Dutch cone penetrometer, in which a cone is dropped or forced into the ground. This is used in sand-bearing strata.

The information gleaned from the borehole in course of boring is logged on a bore hole record form (see Fig. 3.4) and from this a soil profile can be constructed (see Fig. 3.5).

3.4 Laboratory tests

The objective of most tests carried out on soil is to measure the value of its cohesive property and angle of internal friction. That is, the strength by which the soil sticks together (dry sand has no cohesive strength) and the angle at which the sample ruptures,

30

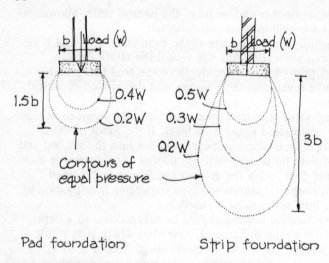

Fig. 3.2 Bulbs of pressure

No of blows	Relative density	Compressive strength
0 - 2	very soft	0 - 24 kN/m²
2 - 4	soft	24 - 49 " "
4 - 8	medium	49 - 98 " "
8 - 15	stiff	98 - 196 " "
15 - 30	very stiff	196 - 392 " "
over 30	hard	over 392 " "

Fig. 3.3 Standard penetration test values for clay soils

indicating the opposition to movement created by the soil particles gripping each other (soft clay has practically no frictional resistance). With these values one can calculate the shear strength and hence the ultimate bearing capacity of the soil.

The most commonly employed laboratory test, the triaxial compression test, is used to determine both these values. This is described more fully in *Advanced building construction*, Vol. 1 Ch. 4, but consists basically of a rubber sheath which contains the sample, inside a clear plastic container full of water. Pressure can be applied to the sample by forcing water into the container and an axial load can be applied to the ends of the sample.

BOREHOLE RECORD

Ground level + 20.74 O.D.   Borehole no. ....16........

| Date | Sample | | Change of strata | | Description of strata |
	Depth	Type	Depth	O.D. level	
			0.40	20.34	Top soil
9.12.83	1.00-1.30	SP(5)			Soft yellow/brown sandy clay
	1.75	D	1.87	18.87	
	2.00-2.30	SP(12)			
	2.60	D			▽ Final water level
	3.20-3.50	SP(20)			Medium brown sand and little mixed gravel
	4.00	U	4.34	16.40	
10.12.83	4.50	D	4.71	16.03	Hard red/brown silty marl
	5.00	W			
	5.20	U			
	6.08-6.38	SP(29)			Compact red/brown sand
11.12.83	7.60	D			
	8.00-8.30	SP(36)	8.32	11.82	
	8.40	U	8.62	12.12	Medium hard brown sandstone
			End of bore		

Key to sample types:
U · Undisturbed sample
D · Disturbed sample
W · Water sample
SP · Standard penetration test
   (number of blows in brackets)

Notes
Water first met at 4.62
Rose to 3.20 in 20 min
Water level at end 2.95

Fig. 3.4 Borehole log

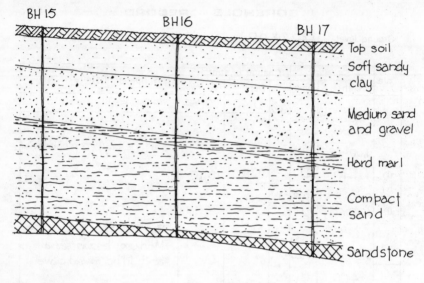

BH 15

BH16

BH 17

Top soil
Soft sandy clay
Medium sand and gravel
Hard marl
Compact sand
Sandstone

Soil Profile 15 to 17   Vert. & horiz. scales 1:100

Fig. 3.5   Soil profile

Another test used is the shear box test, also described in detail in *Advanced building construction*, Vol. 1 Ch. 4. In this, a box comprising an upper and lower half is filled with the sample, which is then compressed by the application of a load to the top of it, and the force necessary to move the top half of the box across the bottom half of the box (thus shearing the sample) is measured.

An alternative objective of laboratory tests is to classify the soil. This does not find its bearing capacity but determines, in the case of non-cohesive soils, the ratio of voids to solid soil particles and, in the case of cohesive soils, the plasticity. The methods used for finding the liquid and plastic limits of clay are given in *Advanced building construction*, Vol. 1 Ch. 4.

All these tests, and others for more specific purposes, such as the test to measure the rate of consolidation of the soil under load, are set out in BS 1377: *Methods of testing soils for civil engineering purposes*.

3.5   Applications

3.5.1   Foundation design

Nearly always the reason that the sub-soil of a site is investigated is to provide information for the foundation designer, but in the course of these tests the facts about the soil itself, its condition, variety, distribution and degree of saturation which are established will be of help to others in both the design team and the building team.

Most soils achieve their shear strength by a mixture of cohesion and frictional resistance (soils are classified as cohesive or non-cohesive but this really refers to a matter of degree rather than to ultimate characteristics), and having measured what these values are their relationship can be established by Coulomb's equation:

$$S = C + p \tan \varnothing$$

where

S = shearing resistance of the soil
C = cohesion of the soil
p = total normal pressure across the shear plane
$\varnothing$ = angle of internal friction of the soil.

Knowing the shear strength of the soil is only part of the information a foundation designer requires to assess its safe bearing capacity; he will also need its density. For a foundation to move it must either compress or displace the soil below it. The sideways movement of soil below the foundation (displacement) must eventually result in an upward movement of the surface, since there is usually nowhere else for the soil to go. Resistance to displacement will, therefore, influence the bearing capacity, and this resistance is partly created by the downward pressure of the soil adjacent to and at foundation level, which opposes the upward movement generated by the displacement. This pressure is the product of the foundation depth and the soil density.

From these tests and analyses, the engineer would find the ultimate bearing capacity of the soil which is then reduced by a load factor to give the safe bearing capacity. This figure represents an upper limit of the stress which could be placed on the ground without causing a collapse, but it does not allow for possible compression of the soil and settlement of the building. Accurate assessment of the amount of settlement which could occur in a building is quite difficult, particularly in cohesive soils. Non-cohesive soils settle very quickly and often have finally consolidated by the time the building is finished, but clay soils can settle slowly over several years.

If the tests, and observation of adjacent buildings, reveal a potential hazard of severe settlement, the safe bearing capacity value would be reduced further to give the 'allowable bearing capacity'.

3.5.2 Trench support

The Building (Safety Health and Welfare) Regulations require that all trenches over 1.2 m deep be supported to prevent the sides collapsing, unless it can be proved that there is no risk in leaving them unsupported. The way in which this support is to be provided and whether any at all is required in trenches less than 1.2 m deep is decided by a number of factors, not the least of which being the type and condition of the soil. Since trench supports can be an expensive item in a contract and must be allowed for in the programming of the work, information of the conditions to be anticipated is of great value to the builder, both when estimating the cost and when planning the execution of the building project.

3.5.3 Ground water control

Many of the difficulties which can arise during the time of excavations on site are attributed to the undesirable presence of ground water. There are many ways of dealing with this, both in the design stage and during construction, and the information on the subject gathered from the site investigation provides a guide to the selection of method to be adopted.

The need to control ground water and the method to be used will affect the cost and duration of the work, and may even influence a decision on the viability of the whole project in the first instance.

3.5.4 Bulking

When a hole is excavated, the volume of the excavated material always exceeds the volume of the hole and due allowance must be made for this in planning a spoil heap on site. The amount of this 'bulking', i.e. increase in volume, depends on the type and nature of the excavated material, for details of which reference would be made to the site investigation. Figure 3.6 shows the amounts by which soils bulk up after excavation.

3.5.5 Embankments and cuttings

The angle at which any discrete material may be cut into or piled up without danger of collapse is its angle of repose, and is approximately the same as the angle of internal friction, as found in laboratory tests. Thus, this particular information is essential for anybody dealing with either an embankment or a cutting for either permanent or temporary purposes. Figure 3.7 gives a guide to what can be anticipated as a safe slope, but reference to test values of the actual material should always be made.

3.5.6 On-site building materials

In the past, many houses were built from stone or clay excavated on site. The clay was moulded into bricks and fired where it was dug out, the resulting hole forming the basement of the property. This is no longer economically feasible but not all excavated materials are carted away even today.

Good top soil is quite valuable and, in most cases, after stripping the oversite the soil is carefully stored or spread as part of the landscaping of the site.

Soil :	After excavation 1m³ becomes:
Chalk	1.6 m³
Clay (soft)	1.2 m³
Clay (stiff)	1.5 m³
Top soil	1.25 m³
Gravel	1.1 m³
Sand	1.05 m³
Rock	1.5 m³

INCREASE IN BULK AFTER EXCAVATION

Fig. 3.6 Table of bulking

Soil	Rake (base to height ratio)
Top soil	1 : 1 to 4 : 1 [a]
Clay (soft)	12 : 1
Clay (stiff)	3 : 1
Chalk	vertical to 0.5 : 1 [b]
Chalk	1 : 1 [c]
Stone rubble	1 : 1
Gravel/sand mix	2.25 : 1 to 1.75 : 1 [a]
Gravel	1.25 : 1
Sand	2.25 : 1 to 1.5 : 1 [a]

Notes:
[a] The variation shows the changes due to moisture content e.g.: Sand (dry) 1.75 : 1
     Sand (moist) 1.5 : 1
     Sand (wet) 2.25 : 1
[b] In cuttings
[c] In embankments

SAFE SLOPES FOR BANKS AND CUTTINGS

Fig. 3.7 Table of safe slopes

If the site investigation reveals clean sharp sand underlying an area to be excavated, this can be re-used as beds to pre-cast pavings where the cleanliness is not critical. Any sand so produced would need to be washed before using for mortar or concrete, which would then make it uneconomical.

Carefully selected excavated material is used for back-filling foundation trenches and the upper section of drain trenches, and where the sub-soil is suitable it can also be used to back-fill round the drainpipe itself instead of granular material brought in. For this latter purpose a test has been devised by the Building Research Establishment to decide the suitability of the excavated material. The first step in this test is to inspect a random selection of the material: if there are any particles more than 38 mm it is unsuitable and generally the maximum size should not exceed 20 mm. The method of test is to fill a tube of 150 mm internal and 250 mm long with the material by pouring without tamping and, when filled, the top is struck off level. This is then emptied and replaced, one quarter at a time. Each of the four layers is thoroughly consolidated with a metal rammer 40 mm in diameter weighing between 1.00 and 1.25 kg with the final surface being brought as level as possible. The difference, (x), between this level and the top of the tube is measured and divided by the height of the tube (H). The ratio x/H is referred to as the compaction fraction.

Any soil material with a fraction of less than 0.1 is suitable; over 0.3 is unsuitable and between these limits it can be used but extra care in compaction is required.

3.5.7 Protection of adjacent property

When working near to adjacent property, a prior knowledge of the sub-soil type and condition is essential if suitable precautions are to be planned to prevent any damage occurring.

Damage in this instance is caused, in the main, by the removal of support by the excavation of either the sub-soil into which the adjacent foundation loads disperse, or the sub-soil above foundation level which resists displacement of the lower layers (see section 3.5.1), or it is caused by disturbance of the ground water condition. The first of these causes can be dealt with by appropriate sheet piling or diaphragm walling techniques, but the second calls for skilful analysis of the existing conditions and the consequences of the proposals. The careful collection and drainage of ground water may solve problems in the new building, but it can dry out the sub-soil downhill from the site with consequential foundation failure, in which case consideration should be given to redistributing the water into the soil on the downhill side.

One point to be guarded against when pumping water from sand (which is often where it is needed) is that as well as the water, significant quantities of sand can be removed, thus eroding the bearing below adjacent foundations with disastrous results.

Chapter 4

Modular co-ordination and performance standards

4.1 The module

Designing buildings with a system of related dimensions derived from a single basic size, or module, has for the last 40 years been receiving an increasing amount of attention and study. It is not, however, a modern invention. The Crystal Palace of 1851 demonstrated the principles of modular co-ordination and, going back further in history to the Greeks ('module' derives from a Latin word 'modus' meaning 'standard of measurement') it can be shown that the dimensions of a temple of this period are all multiples or sub-multiples of a unit equal to 1/60 th of the base diameter of the columns.

These historical examples are really of dimensional co-ordination rather than modular co-ordination because, although a selected dimension is used throughout the design, it is peculiar to each structure and not applied to all building of the time. A nearer approach to the present design philosophy occurs with brick buildings, which are dimensioned to suit the size of a brick and thus the brick becomes the module.

Through the years the discussions have nearly always centred on what the basic module size should be. These have ranged from 8 ft 3 in to a two-stage concept of 14 in and 40 in with a 3 in module making a strong claim, but the choice finally settled on what, at the time, was referred to as the 4 in/10 cm module; now, using SI units, the 100 mm module.

4.2 Preferred and co-ordinated dimensions

Although the use of a 100 mm module will reduce the range of sizes of components to be manufactured and handled on site (this is one of the economic advantages of the system) it still leaves a wide range of choice available, particularly for the larger dimensions in a building, such as the storey height. To achieve further simplification the choices available are reduced to two or three 'preferred dimensions' for either floor to ceiling heights or for storey heights and then other components are co-ordinated to fit with these preferred dimensions. For instance, with a preferred storey height of 2600 mm and floor zones of 200, 250 and 300 mm (see Fig. 4.1) the manufacturers of components which are intended to fit between floor and ceiling only need to produce their units in heights of 2300, 2350 and 2400 mm for them to be universally applicable.

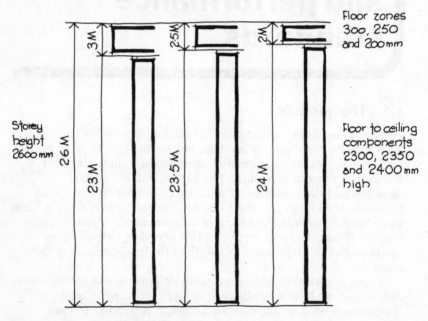

Fig. 4.1 Storey heights and floor zones

4.3 Terminology

Module	A unit of size carefully selected as an increment of dimension. Not related to any specific building size and intended to be universally applicable.
Modular grid	A planning grid used as the basis of building design, normally rectangular,

	with the intervals between the grid lines in multiples of the basic module, but not necessarily regular or constant (see Fig. 4.2).
Basic space	The space allowed between grid planes for the accommodation of a component, including allowances for joints, and other tolerances.
Controlling dimensions	A dimension of the building selected, as a modular multiple, to satisfy a particular requirement which indirectly sets the size of the other components, e.g. floor-to-floor dimensions will determine the staircase rise and the rise of each step.
Component	A manufactured unit, complete in itself, intended to form part of the building and possessing fixed dimensions in at least two directions.
Fit	The relationship between a solid component and the void or space which it is intended to occupy. For example, the size of a door compared to the door frame.
Assembly	The method by which components are connected together whilst retaining the intended distance between the component and the relevant modular or other reference plane.

4.4 On- and off-site operations

Standardisation by either modular or dimensional co-ordination reduces the number of variants to a point where it is economic to produce finished components in large quantities away from the site knowing that when they come to be assembled as a building, no further work should be required. By so doing, all the advantages of factory production, such as a controlled environment, static machinery with high precision and output, continuous work flow, etc., can be employed.

Since any operation carried out under these conditions costs a lot less than the same operation performed under on-site conditions, it follows that the more work that can be put into the component in the factory – the larger and more finished it is – the cheaper will the building system become. The factors which stop this process proceeding to the ultimate point of a complete building made up in the factory are firstly the obvious problems of transport and secondly

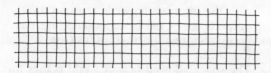

1 M basic modular grid - squares 100 x 100 mm

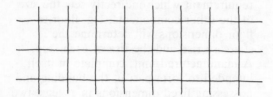

3 M multi-modular grid - squares 300 x 300 mm

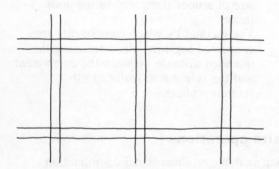

Multi-modular tartan grid   –   main
squares 800 x 800 mm , wall zones 100 mm

Fig. 4.2 Modular grids

the inevitable loss of variety of design. If the component is the whole superstructure then all buildings produced to this system would be identical, but if the components are smaller and can be combined in a variety of ways then the building design options are as numerous as the number of ways the components can be combined.

In designing components for manufacture away from the site, assumptions have to be made about the on-site operations involving their use. These assumptions impose requirements not present with the older system of total on-site production. Not the least of these requirements is the need to set out the on-site part of the work with a degree of accuracy commensurate with that achieved in the factory. It has been found impossible to match the on-site with the off-site precision and so a positional tolerance is always allowed in the component design, but this must not be exceeded when assembling the building. Failure to stay within these tolerances is one of the prime causes of industrialised building failure.

4.5 Performance standards

The design brief for a component will not only be concerned with the dimensional matters described at the beginning of this chapter, but also with the functions the component will be required to carry out. To this end, any dimensional directives will be accompanied by a performance specification. This is a totally different document to the product specification which is frequently written to provide a qualitative description of the building and its parts to complement the drawn indication of the location and assembly of those parts.

A performance specification defines the standards to be attained but does not describe ways of achieving them. For instance, an external wall panel component will, in current practice, have to possess a stipulated degree of thermal insulation. There are many ways by which thermal insulation can be provided and the value of the finished product is the total of the values of the separate parts of the component. Thus the coefficient of thermal transmission of any part derives from the combination of the thermal resistivities of all the materials placed between the interior of the building and the external environment.

Similarly an external wall panel would be required to meet performance standards in relation to weather resistance, thermal movement, sound insulation, fire resistance, accidental damage, vandal attack, aesthetic appearance, handling capabilities, etc.

Unless components are made to a clearly defined performance standard they cannot be included in a product specification. This, as explained above, is a description of the quality of the parts of the building and is the more common meaning of the word 'specification'. In this document the designer's selections are set out in detail but the performance of each element is not described because this was the basis on which the selection was made and so the choice of that component is an indication that it possesses the required characteristics.

Part II

Substructure

Chapter 5

Soil stabilisation

5.1 Purpose of soil stabilisation

The result of a series of on-site and laboratory tests of the properties of the sub-soil of a site is a statement as to whether the sub-soil is capable of sustaining the loads to be applied to it without deformation and, equally important, whether it will retain its measured strength and stability indefinitely. If, as may well turn out, the properties possessed by the soil are inadequate for the designed load, further consideration must be given to the method of design. An alternative or parallel line of action which can be taken is to improve the soil properties in a permanent manner so that the performance requirements are fully or partially met.

Stabilisation may also be considered on a temporary basis to provide the conditions necessary on site to allow the work to be carried out both economically and with safety.

5.2 Importance of moisture content

One of the most significant facts to be extracted from a site investigation report is the moisture content of the soil. This directly affects the strength of the soil. In general, the desirable properties of a soil are improved by a reduction in the moisture content. The converse also applies, particularly in cohesive soils, where an increase in the moisture content is generally accompanied by a deterioration in the strength and bearing capacity. In some soils, again mainly the cohesive soils, removal of soil water will lead to a significant reduction in volume. This shrinkage is not easily confined to the area of work and may cause as many problems as the removal of ground water solves.

An equally important factor is the possible seasonal variations in moisture content which would create changes in the stability of the soil in excess of the amount which the structural design can accept.

Thus, in any soil stabilisation operation, the first consideration is drainage, particularly if the exercise is only required to produce temporary improvement.

5.3 Temporary stabilisation

Where the working conditions are such that stabilisation of the soil is a viable proposition, the methods available are drainage, as explained above; electro-osmosis, which forces the water to drain out; the geotechnical process of freezing; cement grout injection; or sealing with a silica gel.

5.3.1 Drainage

There are many ways by which a site may be drained but generally they are gravity drains or well drains.

Gravity drains can only deal with ground water reasonably near to the surface and consist of a collecting system of porous pipes or rubble-filled trenches connecting to a ditch or similar disposal point. A specialised variation on this is the mole drain, which can only be used in heavy clay, and is formed by drawing a mole plough, consisting of a thin blade with a bullet shaped mole on the bottom end, through the ground to leave a cylindrical drainage channel. These channels lead to a main drain and will retain their shape for quite some time – even years if the surface is not subjected to load.

Well drains collect water in a vertical tube and are either taken down to a permeable stratum through which the water can be drained away or else the collected water is pumped out. A common form of this is the well-point system where the wells are of small diameter, connected via a header pipe to a single pump. By this means the ground water level can be reduced, for as long as the pump is running, by as much as 4.5 m in a single-stage installation.

A multi-stage installation of a second set of well-points sunk at the limit of the excavation made possible by the first set will reduce the water table level still further.

5.3.2 Electro-osmosis

The capillary and frictional resistance forces present in the minute voids in fine-grained soils, such as clays and silts, hold back the flow of the water against the force of gravity and make these cohesive soils difficult to drain. The phenomenon known as electro-osmosis can be employed to overcome this resistance. The method uses two electrodes inserted into the ground between which a small electric current is passed. This causes the water to flow to the cathode which, being made of a perforated hollow tube, collects it ready for disposal, usually by pumping. Given the right combination of conditions this can be a very effective means of dealing with the problem of ground water.

5.3.3 Freezing

This can be applied to all soils and is achieved by circulating low temperature brine through buried pipes to freeze the ground water, thus stopping the flow and temporarily producing a strong sub-soil. It tends to be an expensive method and excavation of the frozen ground can present problems.

5.3.4 Grout injection

Where the results of the soil investigation indicate a material with fairly coarse grain and large voids, consideration can be given to stabilisation by filling the voids with cement/sand grout injected into the ground under pressure.

This, like the method of freezing, prevents the flow of water into an excavation (by filling the voids through which the flow takes place) and also strengthens the soil by restraining movement of the grains.

5.3.5 Silica gel formation

The last method will not work in loose material with a grain size less than 0.5 mm or compact material less than 1.4 mm. In granular materials below these limits, stabilisation, adequate for excavation purposes, can be achieved by the formation of a silica gel within the soil structure. This is achieved by pumping sodium silicate into the ground followed by a solution of a salt, such as calcium chloride, which react with each other to form the required gel.

It tends to be an expensive method because a large number of the perforated boring pipes are required since the pumped solutions will only saturate a cylinder of soil of about 600 mm diameter round each pipe.

The method works similarly to grout injection by strengthening the soil and sealing it against the flow of water.

5.4 Permanent stabilisation

Generally, the permanent improvement of the properties of a soil can only be achieved to a relatively shallow depth (150 mm normally but up to 300 mm with powerful equipment) and thus is more appropriate for the construction of roads, embankments and cuttings rather than for foundations where the stresses penetrate deeply into the soil below.

There are two aims; to produce a soil which will withstand any tendency to deform under load and also which will retain this quality under varying conditions of weather. The main methods used to attain these objectives are: planting trees and grass, compaction, and the addition of binders such as cement, bitumen or resins.

5.4.1 Planting

The binding effect of tree and grass roots growing in embankments and cuttings does not contribute much to the first of the aims of stabilisation – that of resistance to deformation, but it does achieve the second – the retention of the qualities the soil already possesses. Indeed, in some applications it has been necessary to employ this technique not just to retain these qualities, but to retain the soil itself. Mainly this is done to prevent erosion of banks by rain or wind and is used extensively to control sand dunes along the coast.

Small banks thus retained can stand at an angle steeper than the angle of repose, once the roots are established, and to hold the surface until this time it is usual to stretch a net over the surface through which the planting can grow.

5.4.2 Compaction

This is the process whereby the soil particles are packed together by the expulsion of the air formerly filling the voids. It is accomplished by rolling, ramming or vibration.

The removal of air causes a reduction in volume and an increase in the density of the soil. The state of compaction is measured by this last value – the dry density. If a soil is compacted to the point where all the air voids have been removed but none of the moisture has been expelled, it is said to have reached saturation. In practice this is an impossibly high state of compaction for which to aim on site, and so a lower density must be selected as the maximum state of compaction which can be attained.

In considering the maximum state of compaction, the relationship between the moisture content and the dry density must be studied. A wet soil will not contain many air-filled voids and thus only a small amount of compaction is required to approach saturation dry density which, because of the presence of the moisture, is quite low. Conversely, a dry soil has a high theoretical dry density but which, due to the absence of any moisture to provide lubrication, cannot be achieved. This results in a final dry density just as low as when the soil was wet. Between these extremes is an optimum point where the moisture content is low enough to produce a high theoretical dry density but yet high enough to provide the lubrication required to come near to this theoretical value when subjected to a standard degree of compaction.

British Standard 1377 gives methods of testing the compaction of soil and of measuring its moisture content from which the optimum moisture content can be inferred. With this information, decisions can be made, firstly on the adjustments of the natural moisture content which can be made to render compaction effective, and secondly on the compaction equipment which would be suitable for the work.

Tests at the Road Research Laboratory have shown that the smooth-wheeled 8 tonne roller is the most useful compaction equipment, especially for sand and sand/gravel/clay mixtures.

Vibrating plate and vibrating rollers are used for the compaction of soils and are particularly efficient in confined areas. Frog rammers are employed for the compaction of soil in trenches. All of these machines have a low output in any other situations when compared to a roller.

Generally, soil compaction to improve its properties applies to work at the surface, such as the sub-grade for a road, but there is one specialised application of this principle, known as vibroflotation, in which compaction takes place deep in loose granular soil to provide improved foundation bearing capacity. This system uses a vibrator, similar to the vibrating poker employed in concrete work, but much larger, which operates inside a lined bore hole. A quantity of suitable stone chippings or coarse gravel is poured down the hole, the vibrator lowered and the lining slowly withdrawn. The effect is to compact both the placed gravel and the surrounding soil. By repeating the process columns of compacted particles are created which can receive the foundation loads of a building.

5.4.5 Soil-cement stabilisation

More often than not, when reference is made to soil stabilisation it is the soil-cement process which is meant. As the name implies, it is the improvement of the soil properties by the addition of Portland cement to achieve a higher shear strength and a resistance to the ingress of water.

It is mainly confined to the formation of the sub-grade of a lightly trafficked road, or to surface stabilisation to form a temporary road where the naturally occurring soil is of a suitable type and free from any of the soil water salts which react with Portland cement.

Having ascertained from the investigation of the soil that it is suitable for this purpose, the next step is to decide on the necessary moisture content and dry density. A small increase in density makes a relatively large increase in strength for the same cement content and so the dry density should be as high as practicable,

To enable the most suitable proportion of cement to be determined, trial mixtures using say 5, 10 and 15 per cent of cement are made, moulded into sample cylinders, cured and then subjected to an unconfined compression test to find the best one.

5.4.6 Bitumen stabilisation

This method, like the soil-cement method, is most suitable in clean granular soils with very little adhesion. The objective is to coat each grain with a thin film of bitumen, thus binding and waterproofing the soil.

One technique is to use a bitumen emulsion. This is a mixture of bitumen and water in roughly equal proportions which mixes well with sand. Evaporation of the water deposits the bitumen round the sand grains. When the moisture content has dropped to optimum level, compaction by rolling is carried out. This compaction is necessary to develop the bearing value of the soil, since the bitumen emulsion is mainly a waterproofing agent which prevents deterioration of the soil properties by rising capillary action.

For this to be successful the mixture must dry out rapidly, which in this country is unpredictable and usually slow. Assistance in drying out can be given by the addition of a small quantity of Portland cement, but generally other methods of using bitumen are preferred.

The bitumen used in these other methods is 'cut back' by the addition of a solvent or flux of oil. Light petroleum oils produce a rapid curing material, certain paraffins produce a medium curing grade and heavy oils produce a slow curing bitumen. Cut-back bitumen does not readily mix with sand or gravel if it is wet, as is usually the case in this country, and so to facilitate the process about 2 per cent of hydrated lime is added first. This is followed by bitumen, cut back with special road oil which reacts with the lime to form a soap causing a slight emulsification, thereby distributing the bitumen.

5.4.7 On-site processes

The first step in producing a cement or bitumen stabilised sub-grade for a road pavement is to reduce the natural soil to the required transverse and longitudinal levels. These gradients must be constantly checked during the stabilisation procedure to control any possible deviation.

There are three methods by which the cement or bitumen is added to the soil but they are all preceded by the pulverising of the soil. The machinery for this purpose has been developed from agricultural plant and comprises a plough to effect an initial break-up of the soil followed by a rotary tiller to reduce it to a condition where most of it will pass a 5 mm sieve (any large stones are removed). The depth of this pulverisation must exceed the planned depth of stabilisation. The methods used for adding the stabilisation agent are mix-in-place, plant-mix and travel-mix.

Mix-in-place, as the name implies, is a process applied to the soil in its original position, and comprises a preliminary distribution of the cement or bitumen over the surface of the pulverised soil, followed by mixing by a rotary tiller, wetting (in the case of cement stabilisation) and compaction.

The plant-mix method uses a stationary mixing plant to which the pulverised soil and the cement or bitumen are brought. There they are mixed and then returned, spread, levelled and compacted to the desired gradients. This is usually the cheaper of the three methods except where a large area, involving long haul distances, is involved, in which case the travel-mix method would be the most economic.

With the travel-mix method, the pulverised soil from a 6 m wide strip is scraped into a ridge or windrow. If a powdered additive is to be used, this is distributed on top of the windrow but if it is a liquid it is carried on the travel-mix plant. This plant, which comprises a bucket and a pugmill mixer, travels along the cleared strip; the soil is lifted from the windrow by the elevator, mixed and returned to a similar windrow ready for spreading and compaction. This method possesses the advantage of being much less affected by wet weather conditions, as the rain will run off the sides of the windrows without significantly affecting the moisture content of the interior.

Chapter 6

Coffer dams and caissons

6.1 General

Both coffer dams and caissons are structures which are formed to
retain the soil around the excavation of a round, square or
rectangular plan shape. They are mainly used in connection with a
complete building basement or with stanchion or bridge pier bases.

Whilst their purpose is identical, their formation and method of
use is quite different. A coffer dam is constructed in the ground in
its final position round the area to be dug out prior to the main
excavation work. A caisson is constructed in stages at the ground
surface and follows the excavation down into the ground, further
sections being added as the caisson descends. Another distinction is
that a coffer dam, in many cases, is a temporary structure which is
removed after the building work is finished, whereas a caisson is
frequently left in place and incorporated in the building fabric.

The caisson principle used to be employed in the formation of
wells where a labourer at the bottom dug the shaft whilst a
bricklayer, at the top, added successive courses to the brick lining as
it slowly descended, following the excavation. Now, it is mostly used
by civil engineers, rather than builders, because it is ideal for river
working where the first section of the caisson can be formed, towed
into position, sunk and then pumped out to give dry working
conditions for the excavation.

6.2 Situation for use

In normal ground conditions where the soil is reasonably stable and
the level and quantity of ground water does not present a serious
problem, one of the coffer dams described in the next section would
be chosen, especially if the excavated hole is required to be
rectangular. Coffer dams can also be used on wet sites in
conjunction with a dewatering or soil stabilisation process (see
Ch. 5).

These ground control techniques have the effect of temporarily
changing the wet unstable sub-soil to a condition in which a coffer
dam can be used.

The alternative approach to the problems posed by adverse site
conditions is to accept the sub-soil in whatever state it is found and
work through it with a structure strong enough to provide support
and capable of excluding water. This is the principle of the caisson,
of which the several forms are described in section 6.4.

To achieve structural strength without obstructing the central
working area it is usually necessary to make caissons circular in plan,
which rarely coincides with the building plan shape. This fact, plus
the cost of forming a strong circular structure, makes it necessary to
examine the economics of the system closely, although much of the
cost can be offset by using the caisson as the permanent
substructure.

Because the caisson does not rely on any modification of existing
conditions it is often selected for civil engineering contracts where it
is necessary to work through water. In this type of project the
circular plan shape is not disadvantageous, since bridge piers and
similar structures can be designed in a coincidental circular form.

6.3 Coffer dams

There are many forms of coffer dam but fundamentally they are all
designed as retaining walls intended to support the ground adjoining
an excavation, and to exclude most of the ground water. Coffer
dams are not intended to be completely watertight, but they should
restrict any flow to a rate which can be handled by a pump.

The method by which the coffer dam face sheeting achieves its
support strength is determined by the type of ground and the extent
of the excavation (see Fig. 6.1). Where the depth of excavation is
not very great and the ground conditions are suitable, cantilevered
face sheeting, usually sheet steel piling, would be selected.

In this case the sheeting is driven until the depth below the
intended excavation level is sufficient to produce a passive resistance
by the sub-soil great enough to support the upper section of
sheeting. In situations where either the depth of the excavation or

the type of sub-soil makes a cantilevered structure impracticable, a double skin of sheeting filled with sand or hardcore and acting as a gravity retaining structure would be considered. With both these methods, the excavated area is clear of any cross supports or struts. In the circumstances where the presence of cross members will not impede the working, a single skin of sheeting supported by walings

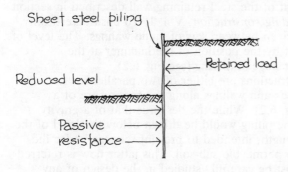

CANTILEVERED FACE SHEETING

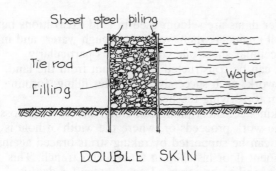

DOUBLE SKIN

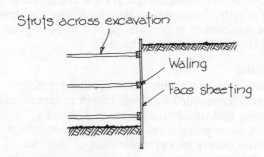

STRUTTED FACE SHEETING

Fig. 6.1 Types of coffer dam

and struts, in a similar manner to trench timbering, would probably afford the most economical solution. Each of these are dealt with in more detail in the rest of this section.

6.3.1 Sheet steel piling

The design and construction of a single sheeted cantilevered coffer dam would follow that of the steel retaining wall described in section 8.6 of *Advanced building construction*, Vol. 1. To ensure verticality the piles would be driven between sets of guide walings. The level of the upper guide is set by the position of the hammer at the completion of the first stage of driving (see Fig. 6.2).

Double skin constructions are either of two parallel rows of sheeting linked by rods and walings along the top, or are of a cellular form (see Fig. 6.2). While the design is that of a gravity retaining structure, the piling would be driven below the level of the filling as a 'cut-off' length, intended to prevent seepage under the filling or through any permeable sub-soil. This latter flow is referred to as 'piping' and must be carefully studied in the design of any coffer dam since its occurrence can rapidly undermine the structure or cause loss of passive resistance, leading to a collapse of the coffer dam.

Double skin coffer dams are seldom used in land excavations but are useful to the civil engineer when working through water, and in this connection the cellular sheet pile coffer dam is particularly attractive because it can be constructed working out from the land, with very little falsework, each successive cell, after filling, forming the working platform for driving the next.

Strutted single skin steel sheet piling can either be braced across the excavation as the work proceeds or, where the width of hole is too great for this, it can be supported by raking struts braced against a strip of the permanent floor installed in a perimeter trench. This leaves a 'dumpling' of soil to be removed once the coffer dam is complete (see Fig. 6.3). Struts across the excavation are prone to buckling due to their length. To overcome this, king piles are driven within the area of the excavation to provide support at each point where the horizontal struts intersect (see Fig. 6.3).

6.3.2 Timber sheeting

Timber was used extensively in the nineteenth century for coffer dams, but is now largely superseded by steel, except in remote areas where timber is plentiful and transport expensive. The main disadvantage of timber sheet piling is the limited depth of cut-off, since it cannot be driven deeply into granular soils or stiff clays without the risk of splitting.

There are several types of timber sheet pile (see Fig. 6.4), all of which are used in a similar manner to steel sheet piling, being driven by a suitable pile driver through a set of upper and lower guide walings (see Fig. 6.2).

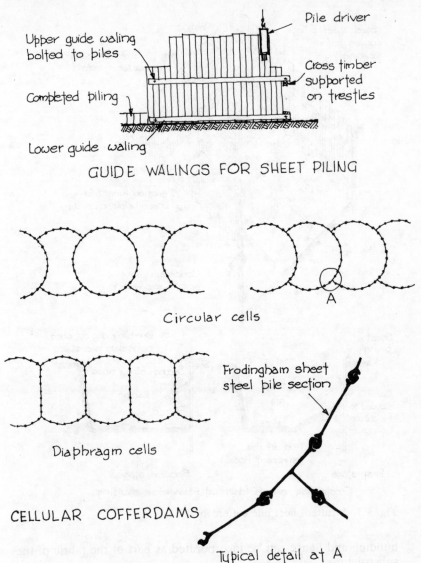

Upper guide waling bolted to piles

Pile driver

Cross timber supported on trestles

Completed piling

Lower guide waling

GUIDE WALINGS FOR SHEET PILING

Circular cells

A

Diaphragm cells

Frodingham sheet steel pile section

CELLULAR COFFERDAMS

Typical detail at A

Fig. 6.2 Steel cofferdams

6.3.3 Precast concrete sheet piling

Generally, coffer dams are intended as temporary structures and the sheet piling is extracted once the permanent substructure work is completed. This is particularly true for steel and timber sheet piles, neither of which will satisfactorily withstand sub-soil conditions for the life of the building. The use of precast concrete sheeting alters this concept in that it has a permanence equal to the life of a

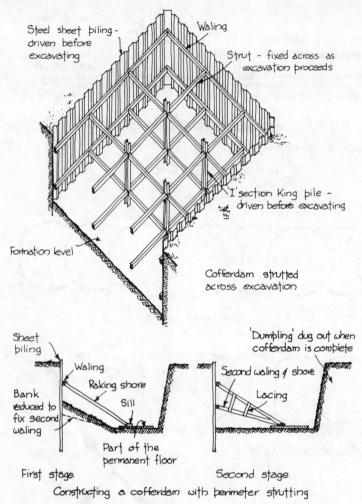

Steel sheet piling - driven before excavating

Waling

Strut - fixed across as excavation proceeds

'I' section King pile - driven before excavating

Formation level

Cofferdam strutted across excavation

Sheet piling

Waling

Raking shore

Sill

Bank reduced to fix second waling

Part of the permanent floor

First stage

'Dumpling' dug out when cofferdam is complete

Second waling & shore

Lacing

Second stage

Constructing a cofferdam with perimeter strutting

Fig. 6.3 Strutted sheet pile cofferdams

building and hence can be incorporated as part of the fabric of the substructure.

In designing concrete sheet piling the same considerations must be applied as are used in the design of bearing piles (see *Advanced Building Construction*, Vol. 4 Ch. 7) in that the reinforcement must be calculated to stiffen the pile against the driving stresses as well as the bending stresses imposed by the retained sub-soil. Each pile is usually fitted with a steel shoe which is inclined, to keep the piles in close contact whilst being driven, and has a joggle joint along its edge to maintain alignment with its neighbour (see Fig. 6.5).

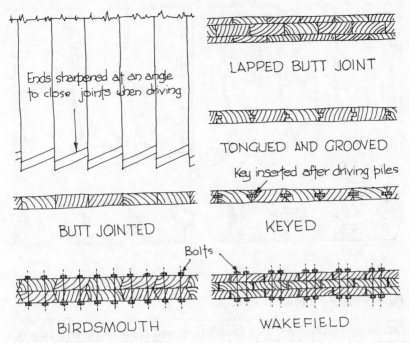

Fig. 6.4 Types of timber pile

6.3.4. Diaphragm walls

These are reinforced concrete walls constructed in advance of the excavation and generally are designed as pure cantilevers, i.e. they are embedded sufficiently far below the excavated level to act as a retaining wall without any other support.

They mainly comprise two forms, either a continuous wall of reinforced concrete, or contiguous piling which is a row of interlocked bored piles. Both these methods are described in *Advanced Building Construction*, Vol. 1 sect. 8.5.2.

6.4 Caissons

As mentioned at the start of this chapter, the distinguishing feature of caissons is that they are constructed above ground, invariably of reinforced concrete, and are sunk to their appointed place in the ground; there to become part of the finished structure. There are four types of caissons currently in use: box caissons, open caissons, monoliths and pneumatic caissons (see Fig. 6.6).

6.4.1 Box caissons

These are mainly confined to river or marine work where the bed

58

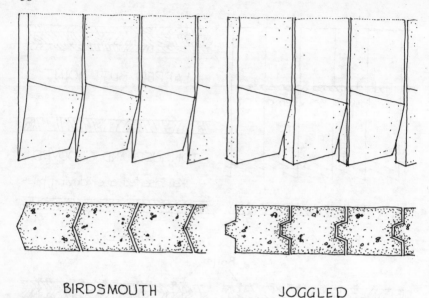

BIRDSMOUTH   JOGGLED

Fig. 6.5 Types of concrete sheet pile

can easily be prepared as a foundation, and consist of a structure of reinforced concrete closed at the bottom and open at the top. After the site has been prepared by dredging and levelling, possibly with a bed or blanket of crushed rock or concrete, the caisson is towed into position, sunk by opening flood valves cast into the concrete cylinder and is either filled with concrete or prepared side panels are knocked out to prevent the caisson from floating.

6.4.2 Open caissons

The open caisson is the most common form and can be used in building as well as in civil engineering work. It consists of a cylinder open at both the top and bottom and with the bottom perimeter formed into a cutting edge.

The first section is cast at ground level or in a shallow starter excavation and then the soil inside it is removed by grab and hand excavation, allowing the caisson to sink. Once the top is just above ground level the formwork, in which the first section was cast, is replaced and a second section cast on top of the first. Excavation then proceeds again and the process is repeated until the desired depth is reached. At this point, concrete is placed in the caisson to form a plug or floor, bonded to the walls, to prevent any further downward movement.

The successive addition of sections progressively increases the weight, which forces the caisson down into the ground. If the type of ground is such that problems with skin friction, preventing the

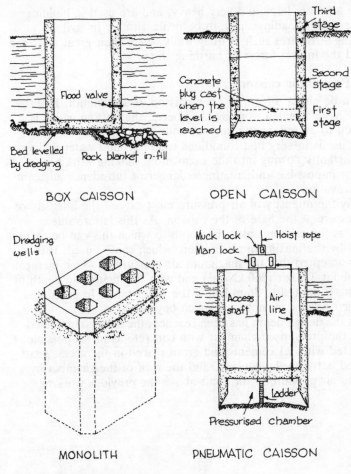

Fig. 6.6 Types of caisson

descent of the caisson, are anticipated, it may be designed so that the upper sections are slightly smaller than the lowest, thus leaving an annular space, which is kept filled with bentonite slurry.

After it has been installed, the caisson may be filled with hardcore, crushed rocks or similar filling and capped with a reinforced concrete slab; or it may be left open for use as a basement if the intended superstructure is heavy enough to prevent flotation of the caisson.

6.4.3. Monoliths

A monolith is really a number of open caissons linked together to form one structure. It is usually made rectangular in plan and the linked caissons form a series of dredging wells.

These are very large and very heavy, and are used in building work where high loadings have to be dealt with, or in civil engineering structures such as quay walls where their great mass can withstand the impact forces of berthing ships.

6.4.4. Pneumatic caissons

These can be either a form of open caisson or a monolith. In either case they have a sealed working chamber at the bottom which is pressurised to exclude water and sub-soil.

Their use is in very bad conditions where ground water or loose soil is constantly flowing into the excavation, making working difficult or impossible and creating a danger of subsidence adjacent to the excavation.

Clearly the intensity of air pressure must exceed the intensity of water pressure at the base of the caisson. As this latter value increases as the caisson sinks, the depth to which this can be carried is limited by the maximum air pressure which can be used. It is generally accepted that the maximum air pressure in which men can work is about 3.5 bar (360 kN/m^2) and with this figure the depth of the sinking is limited to 36 m below the water table. In practice pneumatic caissons are not sunk much beyond 30 m.

When the design depth has been reached, the base is completed by filling the pressurised chamber with concrete as far as possible. It is completed with 1:1 cement-sand grout placed in the access shaft and forced between the concrete and the roof of the chamber by raising the air pressure in the shaft above the previous working pressure.

Chapter 7

Basement construction

7.1 Purpose of basements

In the consideration of the design and construction of any structure, the purpose and function of the various parts must be studied to accomodate the stipulated user requirements. Such is the case with basements. There are two reasons why a basement formation may be included in a building design, and whilst they may both apply they are totally independent of each other, and may even be in conflict.

The first of these reasons is the obvious one that space is required to accommodate activities which economics or practicalities dictate are better sited below ground rather than above. The second is not so obvious and is the result of employing the foundation design technique of balancing the mass of the building against the mass of the soil excavated from the basement to produce conditions of stress in the sub-soil, after building, similar to that which applied before. This is a buoyancy foundation (see *Advanced Building Construction*, Vol. 1 sect. 6.3.4) where this principle is further discussed.

Where it is decided to use a buoyancy foundation it may also be decided to make use of the resulting subterranean space for such purposes as heating plant, car parks, etc. Unfortunately, such purposes demand an unobstructed void, which often conflicts with the structural design requirements of closely spaced cross walls producing a cellular formation. This type of problem is typical of the many which have to be solved between the architect and structural engineer in the very early stages of the building design process.

7.2 Box basements

As the name implies, this is a basement comprising a box set in the ground and generally as such will be subjected to the upward thrusts imposed by the overburden pressure of the adjacent soil. This overburden pressure is the product of the density of the soil, both above and below the water table (below the water table it is referred to as saturated density), multiplied by their respective depths (see Fig. 7.1). This upward pressure will be taken into account when examining the foundation loadings and sub-soil bearing capacities, but it may be incidental to the basement's function of providing additional space.

In all cases, basement constructions must be made strong enough to retain the adjacent sub-soil. This can be achieved by several means, either alone or in combination. Probably the most common solution is to build the basement walls as retaining walls whose toes form part of the basement floor. The installation of this wall requires the soil to be supported whilst the work is carried out, and this can take any of the forms already described in this book, and in *Advanced Building* Construction, Vol. 1 Ch. 8. In Chapter 5 of this book, methods of soil stabilisation or strengthening are described. Some of these are temporary systems which could be employed to reduce the problem of support whilst the basement is constructed, others are permanent and would reduce the loads to be supported by the retaining wall.

Methods of support by sheet piling and the formation of coffer dams are dealt with in *Advanced Building Construction*, Vol. 1 Ch. 8, and in Chapter 6 of this book. Mostly, these are intended to be of a temporary nature but some, particularly the methods using concrete, can usefully be left in place as part of the construction. The sinking of a caisson is a totally different approach but may also be used as a basement construction.

When completed, the mass of the superstructure acting down through the basement retaining walls helps to balance the thrust of the soil, thus assisting the wall in its work. Unfortunately, the retaining wall is usually required to perform its function before the superstructure is built, and thus this assistance is ignored, but the design must allow for the mass to be carried.

The section to be adopted for the wall requires careful study; the major stresses occur at the junction of the stem (the upright part) and the heel and thus the structure needs to be thicker here than anywhere else. The accommodation of this thicknessing can take various forms, as shown in Fig. 7.2. Where the wall is constructed

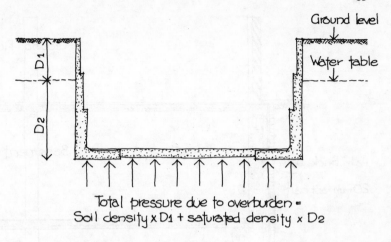

Fig. 7.1 Upward pressure on basement floors

Stepped back   Sloping back   Stepped face

Fig. 7.2 Types of basement retaining wall

close to a boundary, an internally stepped face would have to be
chosen for lack of space outside the wall in which to construct the
other version. The same form would be used where asphalt tanking
is applied to the inside face of the protecting wall (see section 7.4),
but in both cases the thicknessing of the wall reduces the floor area
of the basement. In positions where space and circumstances permit,
a stepped or sloping outer face to the wall would be selected to
avoid this loss of floor area.

In many cases the wall is of normal reinforced concrete, as shown
in Fig. 7.3, but prestressed concrete (see Ch. 10) can be employed
with the advantages of a slimmer section and a degree of resistance
to water penetration, which can make any additional waterproofing
unnecessary.

The basement floor must not only be capable of supporting the
loads to be placed on it but must also withstand the upward

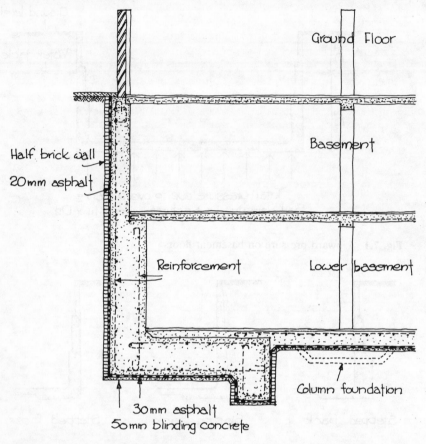

Fig. 7.3 Typical basement detail

pressures explained at the beginning of this section. These upward pressures are dealt with either by making the floor of such a thickness that its mass will resist the thrust, or designing it as an inverted slab bearing upwards against the enclosing or partitioning walls or against the columns of the superstructure.

7.3 Cellular basements or buoyancy rafts

This, as already mentioned, is the form of basement making use of the buoyancy foundation principle. It acquires its cellular character because of the large forces to be dealt with and the need to keep the structure as light as possible to gain maximum benefit from the system. To achieve this lightness the floor, which can be considered as a raft foundation, is interlaced with a regular pattern of stiffening cross walls to produce an egg crate formation.

To remove enough soil to achieve its buoyancy effect, the cellular basement must always be formed deep in the ground. At these depths swelling of the sub-soil on removal of its overburden can present very serious problems. To limit this ground swell, the substructure can be constructed at ground level and sunk to its required level. By omitting the floor of the cellular basement an open caisson of the monolith form results, which is sunk by excavating the cells in a carefully prearranged sequence to leave enough soil within the unexcavated cells to prevent general heaving of the bottom of the excavation. As the level is reached the cells are plugged with concrete to form the floor and the buoyancy raft (see Fig. 7.4).

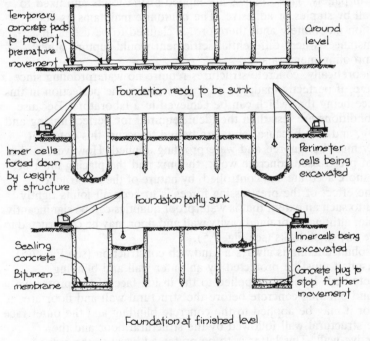

Fig. 7.4 Sinking a buoyancy raft foundation

7.4 Waterproofing methods

The most widely used and strongly recommended method of
waterproofing basements is to form a continuous asphalt membrane
in the walls and the floor but, since this is expensive, alternative
methods exist which are less costly but generally less reliable.
Reliability must be the deciding factor in the choice, as finding and
sealing the source of a leak in a basement can be very time
consuming and expensive.

For very small applications it is possible to construct a basement
inside a heavy gauge polythene 'bag' laid over a blinding bed of
concrete and against protective brick walls around the perimeter. All
joints must be sealed and very great care is called for to avoid any
puncturing of the film. A similar membrane can be formed by a
recent innovation of bentonite gel contained between two skins of
paper in panels. These panels are lapped at the joints and fixed to
the wall by staples or adhesive. The bentonite maintains its gel
condition indefinitely and, therefore, is claimed to be useful in
circumstances where differential settlement would rupture other
waterproofing membranes.

Theoretically, concrete structures require no waterproofing since
concrete, if perfectly made, is itself waterproof; the perfection in this
instance being that which can be achieved in a laboratory. Because
site conditions are less than the ideal required for perfect mixing and
placing, and also because joints are bound to occur, this potential
quality must be ignored and waterproofing applied. However, in the
case of prestressed concrete work, the mix and the placing must be
much more accurate and controlled, by nature of the construction,
and the effect of the prestressing forces is to keep all joints tightly
closed to such an extent that is waterproof qualities can be significantly
reliable, although a drained cavity wall and floor may be constructed to
handle any small leaks (see Fig. 7.5).

Asphalt tanking is always a sandwich construction (see Fig. 7.5).
The asphalt must be protected by an outer wall and blinding
concrete bed. It may be applied to the inside face of the protective
wall and blinding concrete before the structural wall and floor are
cast, or it may be applied to the concrete blinding and the outer face
of the structural wall followed by the structural floor and the
protective wall. This latter is to be preferred because the risk of
damage to the asphalt from reinforcement and vibrating pokers is
greater whilst the structural wall is being formed than when the
protective wall is being built. It does, however, require another
600 mm of excavation, at least, for the asphalter's and bricklayer's
working space. This may not be physically possible because of the
proximity of boundaries and buildings and it certainly will increase
costs.

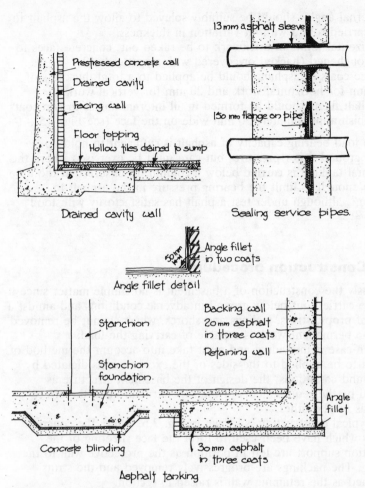

Fig. 7.5 Basement waterproofing

British Standard CP 102: 1973 *Protection of Buildings against Water from the Ground*, makes the following points:

1. Work must be kept dry until all the asphalt is applied and protective coats have set.
2. Horizontal asphalt should be protected by a 50 mm coat of cement/sand mortar.
3. Asphalt must be continuous and, therefore, carried below columns (see Fig. 7.5).
4. Where asphalt is applied to the protective wall, it must be solidly grouted against the structural wall, built afterwards, to prevent inward movement under water pressure.

5. External angles should be suitably splayed to allow the asphalt to be carried round without variation in thickness.
6. Horizontal joints in brickwork to be raked out, concrete faces to be roughened (hacked or covered with a spatterdash).
7. Three coats of asphalt should be applied to a total thickness of 30 mm for horizontal work and 20 mm for vertical work.
8. Asphalt fillets should be formed in all internal angles in two coats to finish approximately 50 mm wide on the face (see Fig. 7.5).

The load bearing capacity of asphalt is sufficient to hold up against ground water pressure, but care must be exercised where the horizontal tanking is carried below a column, to ensure that the base is large enough to limit the bearing pressure in the asphalt to 65 kN/m², although under test asphalt has satisfactorily withstood 130 kN/².

7.5 Construction procedures

Obviously the construction of a basement is no simple matter since it must be carried out below ground in adverse conditions and amidst a forest of props, struts, walings and shores, which cannot be removed until the permanent work is capable of carrying the load.

Each case is individual and must take into account the method of support to be applied to the sides of the excavation as dictated by the ground conditions, the design of the basement structure as dictated by the requirements of the superstructure and the effect each has on the other.

A typical construction is shown in Fig. 7.6, where the concrete walings which have been used to form the face sheeting of the excavation support are left in position as the protective wall to the tanking. The packings are progressively removed and the struts shortened as the retaining wall is raised in section.

7.6 Service entry points

In all buildings, pipes and cables must make their way through the substructure at some point and usually, if tanking has been used, these pipes and cables must pass through this as well, creating a potential source of leaks.

The simplest method of sealing these entry points is by forming an asphalt sleeve round the pipe about 300 mm long before the pipe is fixed and then working the asphalt tanking up to it so that the sleeve and tanking fuse together. This fusion is strengthened by the formation of a two coat fillet around the junction (see Fig. 7.5).

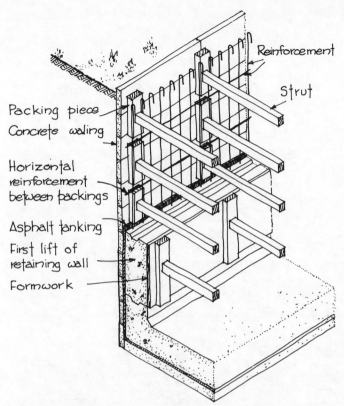

Fig. 7.6 Typical basement construction procedure

In situations where the ground water pressure is high, it is possible for water to be forced between the sleeve and the pipe. To combat this, a metal flange is fitted to the pipe, projecting 150 mm all round, and interleaved between the first and second coats of the asphalt tanking. The flange is either of mild steel welded to the pipe, where possible, or lead sheet sandwiched in the joint of a flanged pipe (see Fig. 7.5).

Part III

Superstructure

Chapter 8

Multi-storey framed buildings

8.1 The selection of the material

Although both possess a high strength/weight ratio, neither timber nor aluminium are strong enough to carry the loads encountered in multi-storey framed buildings and so the choice for the frame must be made from steel; concrete in its various forms of *in situ* or precast reinforced and prestressed; or the more recently developed systems of composite structures of steel and concrete.

Since the material in any of these choices can be made to carry the loads and to maintain stability in all foreseeable conditions such as wind and fire, selection has to be based on a more detailed study of the suitability of the material for the particular design, but above all on cost. The most suitable will generally be the most economic as well.

The proportion of the total cost of a completed building which the structural frame represents can be as low as 10 per cent in, say, an office block with a high standard of enclosing envelope, many internal partitions and a lot of services; rising to 40 per cent in, say, a multi-storey car park with very little other than the frame, floors and lights. Where the proportion is low, differences in cost between one structural system and another will have little effect on the overall cost of the building. But the frame cannot be considered in isolation; design decisions on the structural material and system have considerable effect on all the rest of the work and can influence the final outcome of the building costs to a very great extent. Consider a simple example: taking the span/depth ratio for steel beams as 1:24 and for reinforced concrete beams as 1:20, a 6 m span would require a steel beam 250 mm deep and a concrete beam 300 mm deep. If the clear height from floor to beam is to be maintained the concrete structure, with its deeper beams, would need to be taller – in this case in, say, a 15-storey building 0.75 m taller – with a corresponding increase in both the enclosure of the frame and all vertical elements such as stairs, lifts, pipes, etc.

Apart from the general cost consideration, the final decision on what to use and how to use it will be based on two factors, the design of the building and the process of constructing it.

The building design is the result of the accommodation of the building owner's and the user's requirements within the constraint of the site and legislation, modified by practical considerations of construction. If this results in a simple rectangular frame of standardised elements, steel or pre-cast concrete may be the answer; where irregularity of shapes or non-uniform sections are required, *in situ* reinforced concrete could be best; and where long spans and light loads have to be dealt with, a prestressed system might be considered.

In any building design, the effect of fire must be examined with care and in this respect steel, although non-combustible, does not resist the effects of fire – at 427 to 482 °C (which temperatures are well within those met in building fires), steel loses as much as 80 per cent of its strength. To combat this, fire protection must be provided, often by a concrete casing, which can nullify some of the advantages of the material, such as lightness and speed of erection, but which can also contribute to the strength of the frame, if it is considered as a composite structure.

Limitations of space will dictate the amount of off-site or on-site working which can be undertaken in the construction of the building and hence whether to use a steel or pre-cast concrete frame. These require little space around the building, as opposed to a monolithic *in situ* structure with its attendant demands for concrete production and reinforcement fabrication areas. In most cases, the more work there is to be done on site the slower the rate of progress and the more expensive the result, which fact must be set against the economics of the appropriateness of the system to the building design.

Another subject to receive attention, with respect to the construction of the building, is mechanical plant. Shuttering and reinforcement for *in situ* work is easily man-handled and concrete can be pumped on sites where access is difficult, but beams and columns produced off-site require heavy lifting equipment for which there may not be room but which certainly add to the cost.

8.2 Multi-storey frames

In any multi-storey structure, the loads to be carried arise through the dead weight of the structure itself, the imposed or live load due to occupation of the building, and wind load. The first is a known factor, the second is assumed and can be added to the first to give a constant value for the load on the beams and columns and hence their respective sizes. Wind load is not constant in either magnitude or direction and, rather than imposing direct forces on the beams and columns, tends to cause the building frame to distort by rotation at the beam-to-column joints.

This rotation must be controlled, as it is not practicable to try to eliminate it, and the tall buildings which figure so prominently in our environment sway in the wind to a surprising degree, but it must be restricted to an amount which will not cause distress either to the rest of the building fabric, or to the building occupants! The ways of achieving this are either to stiffen up the joints to make them rigid and thus resistant to the rotation, or to introduce additional elements in the design to provide an overall resistance to the wind load, such as bracing, or a solid core.

8.2.1 Rigid frames

In-situ monolithic reinforced concrete work naturally retains the angle initially formed between the beam and the column no matter what bending occurs in either member and thus is a fully rigid frame, resistant to wind loads without further provisions. This is one of the advantages of the method.

Welding steel beams and stanchions together, although a difficult task up in the air on a building site, does produce a rigid joint; also high strength friction-grip bolted connections are considered to be rigid; thus either will prevent rotation at the frame joints when the wind blows.

Providing rigidity at the joints considerably increases the resistance of the building to lateral loads, but does not prevent distortion of the framing members which must be increased in size to take the additional bending forces (see Fig. 8.1). Where large stresses occur, i.e. in tall buildings, this consequential enlargement of the members makes it an uneconomic solution to the problem and alternatives must be sought.

8.2.2 Braced frames

If, by means of the principle of triangulation, the frame is given an unalterable geometry, the need for the complexity of forming rigid joints disappears. By fixing diagonal braces across selected sets of bays occurring above each other in the frame, stiff planes are formed within the building to which the floors, constructed to act as rigid diaphragms, are attached. The wind pressure on the perimeter walls

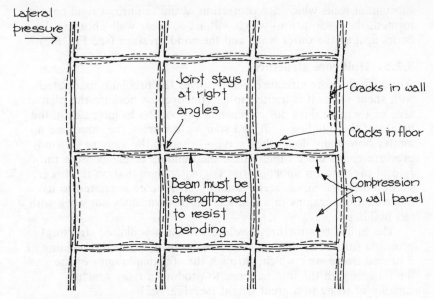

Lateral
pressure
——→

Joint stays
at right
angles

Cracks in wall

Cracks in floor

Beam must be
strengthened
to resist
bending

Compression
in wall panel

Fig. 8.1 Distortion of a rigid frame

is transferred to the floors and from them to the braced bays of the frame which resist any movement (see Fig. 8.2).

8.2.3 Shear wall structure

A shear wall structure solves the lateral pressure problem in the same way as the braced frame, i.e. by forming rigid planes up through the building to hold the floors, but in this case the rigid planes are solid walls of sufficient stiffness to resist the lateral forces (see Fig. 8.2). To achieve the requisite stiffness, these shear walls should be imperforate, which means that within the planning of each floor allowance must be made for solid slabs of wall, one above each other, without any doors or any other openings in them. (A similar plan constraint applies to the braced frame but is not so keenly felt since access is possible through the triangulations of the bracing.) Clearly such an internal constriction is not always acceptable to the building user and should give food for thought to anybody proposing to alter the building at a later date.

8.2.4 Core structure

The plan problem presented by the shear walls is eased if they are associated with the other elements in the building which have a vertical continuity throughout the height of the structure, such as lifts, staircases, main service rises, heating duct rises, etc. If these are grouped centrally in each storey and surrounded by the substantial walls which fire protection would require, a rigid core is formed, the enclosure to which will act as shear walls, bracing the floors against the outer walls and the wind pressure (see Fig. 8.2).

8.2.5 Hull-core structure

Although the core structure neatly solves the problems associated with shear walls, it is limited in its effectiveness because the plan area of the core does not increase in proportion to increases in the height of the building – lift and stair wells remain the same size no matter how many floors they serve – whereas the amount of wind pressure grows more rapidly than the building height because the rate of pressure on upper stories is greater than that on the lower, sheltered floors. Since lateral stability of the core is related to its horizontal dimensions this solution, on its own, does not work with tall buildings.

The hull-core structure develops the shear wall/core structural principle further by adding a rigid outer envelope to the building of a braced framework which, through the stiff diaphragms of the floors, acts with the internal core to produce a rigid structure capable of rising to a great height (see Fig. 8.2).

8.3 Forms of multi-storey construction

The methods used can be distinguished by whether the members of the structure are formed in position or made elsewhere and merely assembled *in situ*. A hybrid version of these, the composite structure, partly pre-formed and partly formed *in situ* can, in the right circumstances, achieve the advantages of each with few of their respective disadvantages.

8.3.1 Monolithic structures

A monolithic structure must be formed in position, unless it is so small that it can be transported whole through the streets, and therefore must be of concrete, either normal reinforced or prestressed.

At the critical stage in its life of being cast, concrete is a plastic material capable of taking up whatever shape is desired. The limitations in this respect do not derive from the material itself but from practical and economic considerations of the formwork against which it is cast.

To form a beam in position requires more preparation than in a fabricating shop – the mould is essentially the same but on site it must be rigidly supported in its correct position in space as well – the placing of the concrete is more expensive (it has to be transported further) – and the progress of the building is hindered by delays whilst the concrete sets and by the obstruction of the forest of

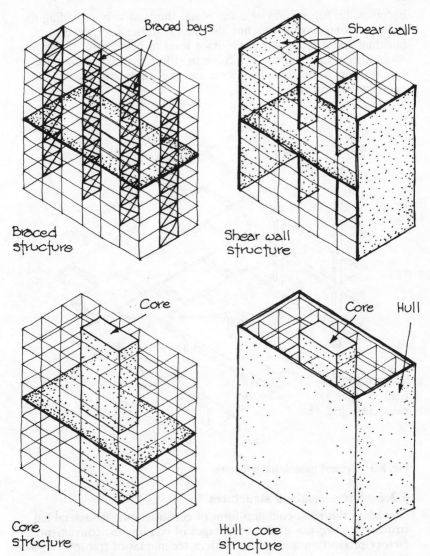

Fig. 8.2 Multi-storey frames

props. With all these disadvantages against it, great use must be made of the plasticity of the material and also of the resulting rigidity of structure. For this reason, monolithic structures can be recognised by their departure from the simple straight beam and column and rectangular connections to shapes which more closely reflect the intensity of stress at each point (see Fig. 8.3).

Being monolithic, the columns, walls, floors and beams all act together to resist the stress imposed on any of them; the floor slab becomes the top section of a tee-beam, thus not only providing the requisite compressive area but also lateral stiffness to prevent buckling and a reduction in the floor level to beam soffit depth; the walls stiffen the columns; the columns stiffen the beams and vice versa, through the rigidity of joint and the whole acts together to resist lateral forces.

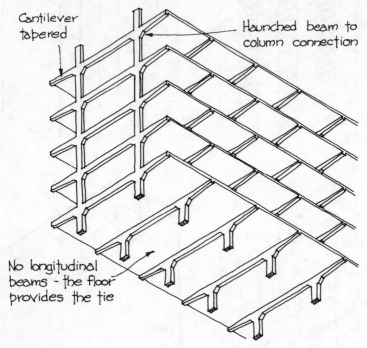

Fig. 8.3 Typical monolithic structure

8.3.2 Site-assembled structures

Nowadays the most common form of construction, site-assembled structures make use of the advantages of continuous, convenient, factory production and utilise modern techniques of transport and handling.

Whether the material used is steel or reinforced concrete depends on the combination of a number of factors already mentioned in this book and in *Advanced Building Construction*, Vol. 1, but the method of erection and assembly is much the same. The sections are hoisted by crane and bolted together. It is in the design of these connections that the engineer's skill lies.

In the case of site-assembled prestressed concrete frames the method of connection is likely to be different, particularly if the members are post-tensioned (see Ch. 10), since the methods of prestressing are also employed to hold the structure together.

8.3.3 Composite structures

In many cases, steel beams are a better choice than concrete beams but, in all but domestic work, concrete floors offer the best choice. In the past, the steel beams have been designed to support both the superimposed floor load and the dead load of the concrete floor without any other assistance. Now, by the use of shear connector studs welded to the top flange of the beam and cast into the floor slab (see Fig. 8.4) the latter acts as lateral stiffening to the steel beam, increasing its resistance to buckling and, usually, reducing the size of member required. This is an example of what is termed 'composite structure'.

Similarly when, previously, a steel column had to be encased in concrete for fire protection, no account was taken of the load-carrying potential of the casing, although in some designs the stanchion size was reduced by virtue of the improvement in the slenderness ratio the casing produced, but with a composite structure the full contribution of both materials is taken into account.

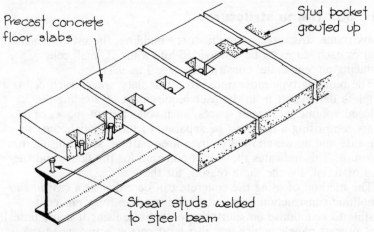

Fig. 8.4 Composite construction

Chapter 9

Concrete panel buildings

9.1 Box frame structures

A box frame structure is a multi-storey building, the use of which requires each storey to be subdivided into a lot of small cells producing an 'egg-crate' construction (see Fig. 9.1).

The building type most suited to this is flats, where each cell is a complete flat. Apart from the user requirement of dividing the enclosed volume into small spaces, each collection of spaces or rooms comprising a flat must be separated from its neighbours on each side and above and below by a fire-resisting, sound-insulating enclosure. This indicates the use of concrete for the floors and the same material, for the same reason, for the walls.

The method of using the concrete can be either as a cast *in situ* monolithic construction (where a repetitive plan form makes it possible to economise on shuttering costs by re-using it many times), or as precast panels, which are also most cost effective when the plan is repeated. These panels are frequently precast in an off-site production area but where special factors apply, such as the particular erection programme, problems of access to the site or remoteness of the site, it is quite feasible to cast the panels on site.

9.2 Structural stability

There are two aspects of structural stability to be considered, firstly the combined effect of dead, live and wind loads and secondly the

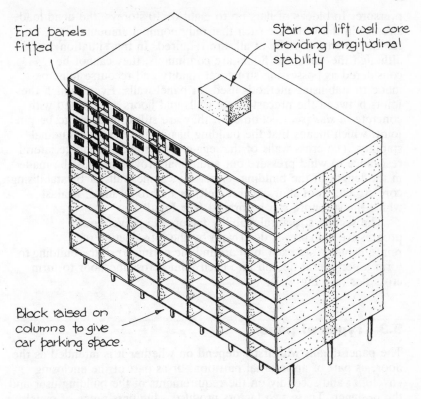

End panels fitted

Stair and lift well core providing longitudinal stability

Block raised on columns to give car parking space.

Fig. 9.1 Egg crate structure

Building Regulation requirement that in a building of five or more storeys the structural failure consequent upon the removal of any structural member shall be restricted to the storey above and below the one being studied. This second factor has particular relevance to precast panel construction; indeed, it was the progressive collapse of the Ronan Point flats, built of panels, which brought about this particular Regulation.

With the small spans involved, domestic loading and the possibility for a cellular plan to have two-way spanning floor slabs, the stresses imposed by the dead and live loads are easily handled by reinforced concrete in any form. It is possible, however, that the thickness required for load carrying purposes would have to be increased to achieve the mass the floor must possess for sound insulating purposes. However, when considering wind load, the difference between monolithic and precast panel structures in the degree of rigidity of the floor-to-wall joint, becomes significant. The monolithic structure, if the walls are reinforced, can be assumed to be a rigid structure as described in the previous chapter and, therefore, inherently resistant to the lateral forces caused by wind pressure. In blocks of flats up to eight or 10 storeys, the dead load stresses in the walls are such that only nominal amounts of reinforcement, or none at all, are required. In this situation, although the walls and floors are continuous, they cannot be considered as possessing structural rigidity and recourse must be made to stabilising methods used for panel walls. Even though the joints between the precast panel walls and floors are formed with concrete *in situ* (see section 9.5), they are still considered to be pin-joint, which means that the building has no inherent longitudinal stability. The cross walls of the 'egg-crate' provide adequate lateral resistance to wind pressure but additional provisions must be made in the length of the building, either by a spine wall, or by stabilising core structures of stair and lift wells or bathroom units, against which the structure can be braced (see Fig. 9.1).

The second structural stability factor, the prevention of progressive collapse, is dealt with by the introduction of reinforcement into each joint around the perimeter of the building to form a peripheral tie and into each joint across the floor to form cross ties (see Fig. 9.3).

9.3 Types of panel

The panel details will firstly depend on whether it is intended as the floor, as part of an internal partition, or as part of the enclosing envelope, and secondly on the requirements of the building user and the designer. These two factors produce a limitless range of panels but yet with certain common details.

Floor panels must be strong enough to carry the superimposed load, usually must provide adequate sound insulation and must have upper and lower faces suitable for the finishes to be applied. The panels may be either solid or cored (see *Advanced Building Construction*, Vol. 1 Fig. 17.1), and although the latter may appear to be the more economical because of the saving in concrete effected by the hollow cores, this may be off-set by the greater depth of panel necessary since they can only span one way, whereas the solid panels can be reinforced as two-way spanning slabs. Furthermore, sound insulation measures would need to be added because the lightness of the cored panel makes it less absorbent of sound energy than the solid slab.

The upper surface of a floor panel must be sufficiently true to receive the finish to be applied (it is most unlikely that the precast surface would be left as the finished floor since it would be aesthetically unacceptable and ineffective against impact sound), it may be directly screeded or, more commonly, it may be overlaid with a sound deadening glass fibre or mineral wool quilt, on which a floating floor of screed or timber is laid.

The floor panel soffits can be produced with a fair face, usually with V-jointed edges, which merely require decorating or they can be keyed ready to receive a plaster finish.

Internal partition panels, like floor panels, can be either solid or hollow cored and, even when load bearing, only need nominal reinforcement, mainly to prevent fracture whilst being handled. If sound insulation is required, as would be the case in a wall separating adjoining flats, the mass of the panel must be checked, particularly the hollow panels, to ensure it meets the Building Regulation requirement of 415 kg/m^2 of face area inclusive of any plaster finish. The panel faces can be finished fair for direct decoration, or prepared to receive an applied finish. Fair-faced panels eliminate a finishing trade and thus shorten the building programme, but panel joints will be left showing, which may not be acceptable. Great care in handling is also required to avoid damage to the faces or, more particularly, the edges of the panels.

External wall panels must not only possess the strength necessary to resist the dead, live and wind loads but must also incorporate adequate thermal insulation and condensation control. To be truly economic, the panels would also have an external face finished to the required standards of aesthetic and weather resisting standards, and an internal face either finished fair or prepared in the same way as the internal wall panels.

In many cases, external panels are produced as a sandwich of two concrete leaves with sheets of expanded polystyrene board between. The inner leaf of the panel is designed to be load bearing and to carry the outer leaf which provides the weathering and external finish. The connection between these two leaves must be carefully designed to be strong enough to hold them together, but not so rigid that the greater thermal and moisture which occurs in the external leaf does not warp the internal one.

9.4 Panel production

Whether the panels are cast in a factory or on site, the manner by which this is done must be related to the details of the panel itself and the way in which it will be handled.

A shallow tray which can be filled to the top with the concrete mix and left to set would seem to present a simple logical way of casting large, relatively thin panels, and for many applications it is used but, where appropriate, there are distinct advantages in casting panels vertically.

Whenever the panel has complications or difficulties it is cast horizontally, thus external panels are usually made in this way, starting with a lining or layer of facing material laid in the bottom of the mould, followed by the concrete for the outer leaf with the necessary ties cast in, followed by the expanded polystyrene sheets and finally the inner concrete leaf, the surface of which is then trowelled up to the required finish. If the external finish is a worked texture the panel would be cast the other way up. See *Advanced Building Construction*, Vol. 1 Ch. 16 for details of finishes on concrete. Horizontal casting would also be used for any panels, internal or external, where openings have to be formed in specific positions for doors, windows, service access points etc.

Having cast the panel it must then be left to cure and there is the first of the disadvantages of the horizontal method; a considerable area must be set aside for up to 14 days where these large panels can be left undisturbed. Once they have cured they must be lifted out of the mould and stacked ready for transportation to their destined position. This lift from longitudinal to vertical imposes bending forces on the panel which, if it is a wall panel, will never occur when it is fixed but yet tensile reinforcement must be provided otherwise the panel will crack. The cost of having to insert reinforcement which subsequently is redundant is another disadvantage of horizontal casting.

By casting wall panels vertically, they are already in the attitude they will occupy in the building and thus obviate the need for any tensile reinforcement. It is also possible to produce panels with a better fair faced finish both sides by this method since both faces are cast against a mould – which is always superior to a hand trowelled surface. Vertical casting also takes up less floor space, but is restricted to simple panels requiring no finishing work or any special features incorporated.

9.5 Jointing

As with all prefabricated building systems, the success of the construction depends on the skill with which the joints have been designed and carried out. In a precast concrete panel building there are three types of joint to be studied; the edge-to-edge joint between internal panels; the edge-to-edge joint between external panels and the floor-to-internal and floor-to-external wall joints. The first of these would be a simple butt or joggled joint between the internal panels, to be completed by grouting or pointing, but the other joints require more careful consideration.

Edge-to-edge joints in external panels must be capable of keeping out the rain, even when subjected to high wind pressure; they should offer the same degree of thermal insulation as the panels to avoid pattern staining; they must accommodate thermal and moisture movement, and they must be able to achieve these objectives within the variation of joint width which the manufacturing and positional tolerances will allow.

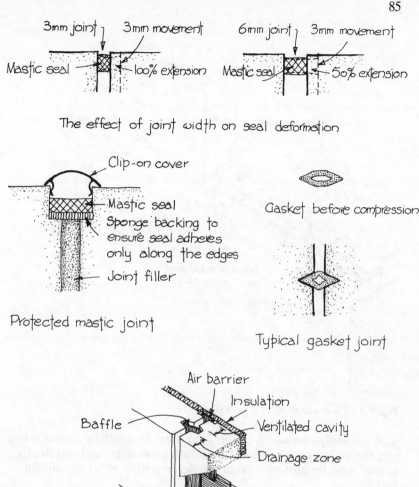

3mm joint | 3mm movement

Mastic seal — 100% extension

6mm joint | 3mm movement

Mastic seal — 50% extension

The effect of joint width on seal deformation

Clip-on cover

Mastic seal
Sponge backing to
ensure seal adheres
only along the edges
Joint filler

Protected mastic joint

Gasket before compression

Typical gasket joint

Air barrier
Insulation
Baffle
Ventilated cavity
Drainage zone
Lead flashing

Typical drained joint

Fig. 9.2 Concrete panel joints

The requirements can be met simply by stopping up the joint
with mastic but, to ensure durability of this, which is referred to as a
closed joint, three points in particular require care. Firstly, since the
joint relies on the adhesion between the mastic and panel edge, the
latter must be sound, clean and dry at the time of application of the

86

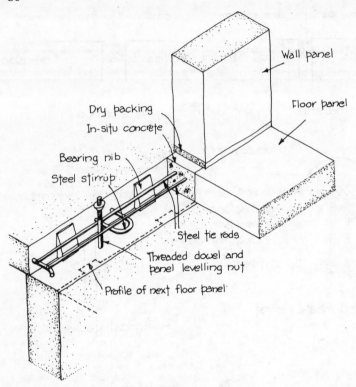

Fig. 9.3 Concrete floor and wall panel joint

seal; secondly, the width of the joint must be carefully controlled to suit the ability of the mastic to tolerate movement; and thirdly, the mastic must be protected from the deteriorating effect of sunlight and air.

The first is a question of site control to ensure that the panel edges are not damaged or dirtied either in storage or while being handled. The last point, exposure of the mastic, should be covered by the panel designer, usually by means of a clip-on cover plate (see Fig. 9.2) but the width of the joint is subject to a high degree of dimensional accuracy on site, and is much more critical than is generally appreciated.

Both extremities of joint width must be controlled. If the joint is too wide the mastic cannot bridge it and will fail by sagging; if it is narrow the degree of movement or deformation of the mastic increases considerably; for example, if the anticipated movement across a joint is 3 mm and the minimum joint width is also 3 mm, at the point of maximum movement the joint becomes 6 mm wide and the mastic filling has been stretched to twice its size or a 100 per cent extension (see Fig. 9.2). If, however, the minimum joint had been 6 mm, then the maximum would have been 9 mm and the extension of the mastic only 50 per cent.

The mastic-filled joint is described as a 'soft-seal'. An alternative to this for a closed joint is a 'compressive seal' achieved by a gasket. In this case (see Fig. 9.2), the maximum anticipated joint width must always be less than the original size of the gasket so that it is always under compression.

A different approach to the design of the joint is a drained joint, in which the water is allowed to enter the joint where it is collected and drained away before it can reach the seal and insulation placed at the back of the joint (see Fig. 9.2).

The structural connection between the floors and the walls is usually effected by the use of *in situ* concrete as shown in Fig. 9.3. The edges of the floor slabs have nibs cast on so that they can gain a bearing on the lower panel but yet leave room for the concrete filling. The upper panel is correctly positioned before the filling is placed by the adjustment of the levelling nuts on a threaded dowel rod cast into the top of the lower panel. The joint itself contains steel reinforcing rods which act as ties to hold the structure together at each storey level, thus preventing the progressive collapse mentioned earlier in this chapter.

Chapter 10

Prestressed concrete

10.1 Principle of prestressing

Normal reinforced concrete is unable to realise fully the potential of the materials used, for two reasons. Firstly, simple beam design allows for all the compressive stress, due to the dead load of the beam and the applied load, to be absorbed by the concrete between the top face of the member and the neutral axis; and the tensile stress to be contained by the reinforcement near the bottom face. This is because concrete is so weak in tension that its value is ignored, and it means that the only function performed by the concrete of the beam below the neutral axis is to connect the steel to the upper part of the section. The stress distribution diagram for normal reinforced concrete is shown in Fig. 10.1.

The second reason why the potential of steel and concrete is not fully employed is that under load the materials deform – the steel lengthens and the concrete shortens – the result that the beam bends and tension is induced in the concrete at the bottom, which then cracks. These cracks do not mean that the beam is about to fail but they can expose the reinforcement to corrosion and consequent weakening, and thus must be restricted to a maximum width of 0.3 mm as recommended in BS 8110: 1985. To achieve this, the stresses in the materials must be restricted to limit the deformation. Therefore neither the high tensile strength possible in steel nor the large compressive value of concrete produced by modern techniques can be fully employed.

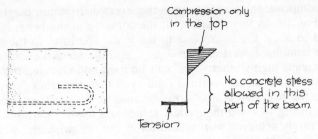

Compression only in the top

No concrete stress allowed in this part of the beam.

Tension

Normal reinforced concrete beam.

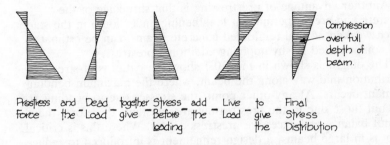

Compression over full depth of beam.

Prestress and Dead together Stress add Live to Final
Force - the - Load - give - Before - the - Load - give - Stress
loading the Distribution

Stress development in a prestressed beam.

Fig. 10.1 RC and prestressed beam stress distribution

By prestressing the beam, that is by applying forces to the beam before the working load is applied to counterbalance that load, these disadvantages are overcome and smaller beam sizes capable of spanning greater lengths become possible. This force which is induced before the load is applied – hence 'pre'-stress – is brought about by stretching the tensile reinforcement by a predetermined amount and then anchoring the ends of the steel members to the ends of the concrete beam.

The steel, in an endeavour to return to its original length, exerts a compressive force on the concrete at the bottom of the beam. This force, when combined with the forces due to the self-weight of the beam, produces a stress distribution in the concrete consisting of high compression at the bottom face and nothing at the top face. In some designs a small tensile force is allowed to develop at the top which is permissible because the high-grade concrete used has an appreciable tensile value (see Fig. 10.1).

As the load is applied, the stress distribution reverses, the normal compression which occurs in the top of the beam overcomes the tension due to the prestress and converts it into a compressive stress, and the normal tension in the bottom of the beam reduces the prestress compression to zero to give the stress distribution diagram shown in Fig. 10. In this situation all the concrete of the beam is being used to resist the forces due to the load. The force in the steel due to the combined prestress tensioning, and the tension created by the load, means high-strength steel can be used and cracking only occurs in the event of an overload (and even then the cracks close again once the overload is taken off, because of the prestress). What are shown in Fig 10.1 are theoretical values and loading patterns which are rarely achieved in practical terms but the principle, when applied in the right combination of conditions, can achieve worthwhile savings in materials even if the practice falls short of the theory.

Another advantage of prestressing is that shrinkage of the concrete and creep (permanent lengthening under load) in the steel, which weaken normal reinforced concrete beams, can be estimated and compensated for by applying additional prestress.

The diagrams shown in Fig. 10.1 show the patterns of stress distribution mid-way along the beam, where the maximum bending moment occurs. At any other point, the stresses due to bending are less but those due to the prestressing are constant; therefore, they do not balance and excessive prestress occurs. Where this is critical, such as in large beams, a design refinement is introduced to reduce the prestress value to correspond roughly to the bending moment value. This is done by adjusting the distance between the neutral axis and the tensioned steel, since this dimension is part of the lever arm of the beam's moment of resistance and determines the effect the prestressing force has on the beam section – the smaller the dimension the less the effect. This can be done either by changing the shape of the beam so as to bring the neutral axis nearer to the straight line of the steel or by keeping the beam section constant and curving the steel up towards the axis at the ends (see Fig. 10.2).

The first of these leads to increased formwork costs, problems with shear forces where the section is reduced towards the bearing and possible difficulties with fitting the pitched or curved upper face to the floors of the building. Curving the steel upwards provides additional shear resistance but prevents the fabrication of a series of such beams in a long line in a factory as described in section 10.3, nor can it be done if the beam is to be pre-tensioned as explained in the next section.

10.2 Materials

Because the method exploits the maximum potential strength of the contributory materials, increased economy of section can be achieved by using high grade components. For structures where the prestressing force is applied after the concrete has set and cured

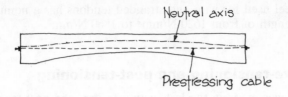

Beam with arched soffit

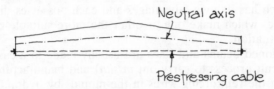

Beam with pitched top face

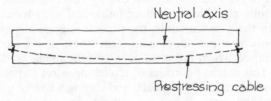

Beam with curved cables

Fig. 10.2 Adjustment of the prestressed force

(post-tensioned) a mix giving a 28 day crushing strength of 30 N/mm² is used and for work where the prestress is applied before the concrete sets (pre-tensioned) 40 N/mm² concrete is used.

The reinforcement used is also of a higher grade than for normal reinforced concrete which uses mild steel. In prestressed work the steel is cold drawn high-carbon steel conforming to BS 5896: 1980 in the form of wires ranging in size from 4 to 7 mm diameter and stranded tendons made up of a core wire round which are wound a number of wire strands to produce reinforcement looking rather like wire rope. The small diameter wires are convenient to handle and store, as they are kept coiled on a drum and a multiplicity of small diameter wires instead of a few larger rods offers a greater bond strength, because of the increased total surface area of the steel. High-tensile steel wires can be obtained in a plain round form but often they are crimped or indented to improve the strength of the bond. Stranded tendons, like the small wires, are also stored on a drum, and, being flexible, are easy to handle. They are made up with seven or 12 strands and are capable of taking very large prestress forces.

The steel used for wire and stranded tendons has a nominal tensile strength of from 1620 N/mm² to 1860 N/mm².

10.3   Pre-tensioning and post-tensioning

Reference was made in the last section to pre- and post-tensioning. They are both forms of prestressing concrete and define the method of manufacture and the point in time when the prestressing force is applied. Each has its own advantages and each possesses distinct characteristics which make one or other the more suitable choice in any given situation.

Pre-tensioned beams and slabs are usually made in a casting shop (it is a difficult process to carry out *in situ*) and manufacture consists of stretching the reinforcing wires in the moulds by hydraulic jacks operating against end abutments fixed to the casting shop floor or against the ends of the moulds, which must be correspondingly strengthened to take this force, and casting concrete round them. Whilst it is quite possible to produce individual units formed and prestressed in their own moulds, the process is generally employed in connection with a prefabricated system where a series of identical units are required which are most economically made by 'long line' production. This method consists of stretching wires between two anchorages, as much as 120 m apart, and casting a row of units, end to end, round the wires. When the concrete has set and cured, the jacks are released and the wires are cut to separate the units (see Fig. 10.3).

Post-tensioning is nearly always carried out on site because the whole structure can then be prestressed rather than the individual conventionally joined elements. The method is to cast unreinforced concrete units with ducts in them; when cured they are positioned end to end, prestressing tendons are threaded through the ducts, anchored to one end of the beam and tensioned by jacks bearing against the other end. When the full prestress load has been developed the tendon is anchored at the jacking end and the jack removed (see Fig. 10.3).

One of the advantages of post-tensioning is, as already mentioned, that the whole structure can be prestressed. Thus a pre-cast concrete structure, particularly a panel system, can be made to act as a monolithic structure and so enjoy the benefits of a rigid frame while retaining the economies of off-site prefabrication. Another problem solved by post-tensioning is the transportation and handling of pre-cast units. These usually are very large, heavy and awkward but if a prestressed post-tensioned system is used, the structural members can be made up of a row of smaller blocks ranged end to end, as already shown in Fig. 10.3. Small blocks are obviously much easier to transport and handle, and the savings thus

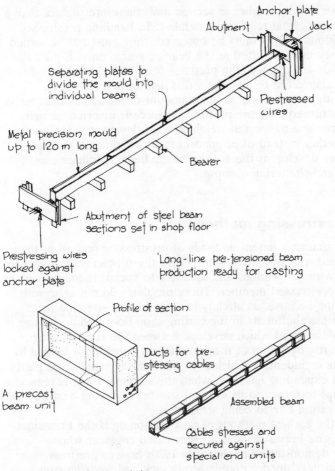

Anchor plate

Abutment

Jack

Separating plates to
divide the mould into
individual beams

Prestressed
wires

Metal precision mould
up to 120 m long

Bearer

Abutment of steel beam
sections set in shop floor

Prestressing wires
locked against
anchor plate

'Long-line pre-tensioned beam
production ready for casting'

Profile of section

Ducts for pre-
stressing cables

A precast
beam unit

Assembled beam

Cables stressed and
secured against
special end units

Pre-cast post-tensioned beam.

Fig. 10.3 Pre-tensioned and post-tensioned beams

effected can off-set the increased cost of the higher grade materials
and the additional work involved in assembling and tensioning the
beam. A third advantage of post-tensioning is that the steel can be
curved up in the beams, thus maintaining the balance between the
prestress forces and those due to the applied load, as already
explained in Section 10.1.

Pre-tensioned members enjoy the same advantages as normal pre-
cast concrete: efficiency of production and maintenance of quality in
factory controlled conditions, economical use of moulds etc., by
repetition and standardisation, greater speed on site, no shuttering
obstructions, etc.; and the same disadvantages of transporting and
handling large units, but with the improvement that prestressed

members are usually smaller in section and, therefore, lighter than a corresponding normal reinforced member. In handling prestressed beams additional points must be observed: they must not be turned over and they must be lifted at the bearing points only. Before the building loads are applied, the prestress force and the self weight of the beam balance out to give the stress distribution shown in Fig. 10.1 with zero stress or a small amount of tension at the top. If the beam is turned over the prestressing force is inverted as well, but the forces due to the self weight remain downwards and so the two act together instead of in opposition with the result that large tensile forces develop at the bottom of the beam, producing an immediate and shattering collapse.

10.4 Prestressing methods

There are numerous patent methods of prestressing and all of them are concerned with post-tensioning where the problems are firstly to stretch the wires or tendons and secondly to anchor them against the end of the prestressed member. These problems do not arise with pre-tensioning because, as already mentioned, the stressing jacks operate against abutments in the casting shop floor and anchorage is achieved by the bond which develops between the steel and the concrete, partly by virtue of the two adhering together – helped by the crimping or indenting which is worked on the steel – and partly by the steel expanding laterally, when the stressing force is removed, in an attempt to return to its original size (after having been reduced by the elongation due to being stressed).

One of the earliest methods of post-tensioning is the Freyssinet system. Eugene Freyssinet was a French civil engineer who successfully demonstrated in the early 1920s how to prestress concrete and developed a double-acting jack and wedging cone method of anchorage. In this, the individual strands of the prestressing tendon are attached to the stressing piston at the back end of the jack which bears against a reinforced concrete anchor cylinder cast into the end of the beam. Operation of this jack then elongates the prestressing tendon by the calculated amount required. When the required prestress has been developed, the second jacking piston is brought into operation. This thrusts against the stressing piston and forces a wedging cone into the anchor cylinder inside the strands of the prestressing tendon. The wedging cone is externally fluted to correspond to the number of strands in the tendon and spreads and clamps the strands firmly against the tapered core of the anchor cylinder. When the cone is firmly in place the jacking stress is released, the tendon strands are cut back, the recess in the beam end filled with mortar and the cable grouted solid through a hole provided in the cone.

10.5 Uses of prestressing

The whole theory and development of prestressing is concerned with reinforced concrete members which are subject to tensile forces. Therefore, its use is mainly confined to beams and floor slabs. Columns carrying compressive loads do not benefit from prestressing unless there is a tendency for them to be bent by lateral loadings, beam eccentricities or counter movements in a monolithic structure.

Concrete shell roofs (see Ch. 12) are sometimes prestressed, partly to deal with the forces present, but largely to allow the shell to be pre-cast in panels which are locked together by the prestressing force. Another application is in concrete retaining walls (see Fig. 10.4 and Ch. 17) in which case the system is employed not only to counterbalance the forces due to the retained material but also because the high grade concrete and the elimination of any cracks by the prestress force produces a virtually waterproof structure which often will not require the expensive process of tanking.

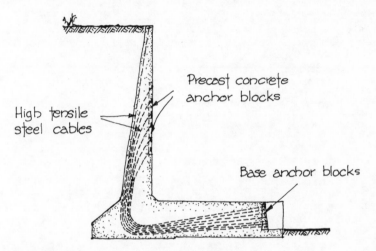

Fig. 10.4 Prestressed concrete retaining wall

Beams which are continuous over several bearings produce complications by the inversion of the stress distribution pattern at the bearing points (tension at the top and compression at the bottom) and while these can be solved in prestressed beams, the difficulties of doing so make the system less attractive economically. In general, prestressed concrete comes into its own on economic grounds with long span lightly loaded structures. With spans below 6 m normal reinforced concrete is usually the cheapest (except for standardised prefabricated individual elements like prestressed concrete plank lintels and some hollow beam and rib and pot floors); between 6 and 9 m spans, factors other than the distance between supports will determine whether normal or prestressed reinforced concrete is the cheapest (indeed the composite form of structure mentioned in Chapter 8 may well prove more economic than either); but over 9 m prestressing becomes worthwhile.

Chapter 11

Crosswall construction

11.1 Principles

The box frame structure described in Chapter 9 is an example of crosswall construction, in that case carried out in pre-cast concrete panels in a multi-storey building. Generally, the term crosswall construction is used in connection only with low-rise buildings such that the plan is appropriate for this system. A suitable building plan is one with a series of walls running from front to back of the building block at regular, fairly close intervals along its length and coincidental in each storey. These walls should also be required to be better than light partitions (e.g., must be fire resisting or sound insulating) so that they must be substantial and therefore, without any additional cost, can be load-bearing.

The most widespread use of crosswall construction is in domestic work. It is ideal for terrace housing where the crosswalls form the party walls between the houses and therefore are at regular intervals in the same position on each floor and must resist the passage of fire and sound. For the same reasons, the use of this system is extended to low-rise flats and maisonnettes. Other building types which lend themselves to this system are hotels, small office blocks, small factory units (although these are often only single storey) and some educational buildings such as classrooms or lecture rooms.

With any repetitive structure, the aim is to make the repeats identical and in that way the building elements can be standardised. For this reason, the use of crosswall construction restricts the planning of the building in that all the crosswalls must be of the same spacing for maximum economic benefit. Thus in any of the uses mentioned in the last paragraph the spaces enclosed (and therefore each house, flat, maisonnette, hotel room, office or classroom) must be the same size. Subdivisions can be made but the main walls must be regular.

The exact dimension between the crosswalls varies from one building to another, but generally it is in the range of 3.5 to 5.5 m with the maximum economy slightly below the middle of this range.

11.2 Structural design

BS 5628: Part 1: 1985 *Structural use of unreinforced masonry* is used as the basis for calculating the loading and stresses in the crosswalls where these are built of bricks or blocks. The subject of calculated brickwork is dealt with in Ch. 14 of *Advanced Building Construction*, Vol. 1. As well as bricks or blocks the crosswalls may also be composed of plain or no-fines concrete as described in the next section.

This structural system, consisting as it does of vertical and horizontal intersecting planes, is much more stable in the direction of the planes than at right angles to them. With the type of construction here being considered the joint between the floors and the walls cannot be taken to have any rigidity as would be the case with a cast *in situ* monolithic construction, nor do the walls have any tensile strength to resist bending, therefore longitudinal bracing of the whole building must be provided to resist wind pressure on the ends. There are several ways by which this can be achieved and which can be used on their own or in combination. They mainly consist either of looking for, and using, walls which run longitudinally or of stiffening the structure by the use of some *in situ* reinforced concrete walls and floors or longitudinal beams. This stiffening tends to defeat the simplicity of the system which is one of its main advantages.

In many cases, and especially in domestic buildings, the space enclosed between the crosswalls requires to be subdivided. If these partitions are suitably built and bonded to the crosswalls they become shear walls (see section 8.2.3) and can readily provide all the longitudinal bracing needed. Where no internal divisions are required a core structure (see section 8.2.4 and Fig. 9.1) of stair or lift wells may be used against which to brace the floors.

Longitudinal stability can be obtained from the infill panels between the ends of the crosswalls, provided they are solidly built and bonded to the crosswalls, but these are always perforated by window openings and thereby weakened. Building solid end panels fails to take advantage of the opportunity the system offers in prefabrication. By returning the ends of the crosswalls to form a T in plan (see Fig. 11.1) sufficient of the stabilising effect of the solid end panels can be obtained whilst still leaving enough to make prefabrication worthwhile. At the same time the cross member of the tee will bring about a separation of adjacent door and window openings which will help to reduce transmission of sound between adjoining houses."

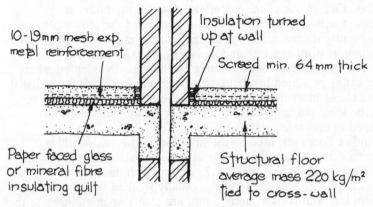

10-19mm mesh exp. metal reinforcement

Insulation turned up at wall

Screed min. 64mm thick

Paper faced glass or mineral fibre insulating quilt

Structural floor average mass 220 kg/m² tied to cross-wall

Fig. 11.1 Concrete floor to crosswall junction

11.3 Materials

Because of their simplicity and familiarity, bricks or blocks are frequently used for the crosswalls but plain *in situ* concrete walls are also suitable; but to compete with the others in cost, the contract must be sufficiently large and standardised to allow the extensive re-use of formwork and a crane to handle it, and the concreting materials.

If bricks or blocks are used, the wall can be either a solid construction or a cavity wall. Neither possesses advantages over the other but in either case, if it is a party wall, it must comply with the Building Regulation requirements as regards its mass for sound insulation and its fire resistance.

Concrete walls can be of either normal dense concrete or no-fines concrete. Dense concrete does not require reinforcement for normal purposes in this situation, but certain selected walls may be reinforced to provide a rigidity for purposes of lateral stability. Even in an unreinforced state, dense concrete walls must be made thicker than required by domestic loads usually encountered in this form of construction, for reasons of sound insulation (see section 11.4).

No-fines concrete is a mix of coarse aggregate and cement without the addition of fine aggregate, as a result of which it contains a large number of small cavities or interstices evenly distributed throughout the wall. The aggregate must be regular in shape, spherical or cubical, and careful control of the amount of mixing water must be exercised to restrict the cement to a coating on the aggregate only and not a filling of the interstices. Having a high void content, no-fines concrete is about two-thirds the mass of dense concrete; being also weaker it means that crosswalls in no-fines concrete must be thicker than in dense concrete, both for load bearing and for sound insulation purposes.

The efficient use of the materials of which the floors are composed will have a bearing on the spacing of the crosswalls and consequently on the economic design balance mentioned earlier in this chapter. Both concrete and timber floors can be used in crosswall construction but the latter are not usually adopted where fire resisting or sound insulating standards are imposed, for whilst it is possible to meet these in timber by the addition of special constructions, the standards are far more easily and satisfactorily achieved in concrete.

Timber floors are used as the first floor in houses of single occupancy and as the upper floor in a maisonnette. The economic span for timber joists is about 4.75 m, at which distance a 50 × 225 joist is required. The deepest joist usually stocked by merchants is 225 mm but widths of 63 and 75 mm are available in addition to the 50 mm. However, these thicker joists represent increases in timber content over a 50 mm joist of 26 per cent and 50 per cent respectively, with permissible increases in the span of 9.5 per cent (to 5.2 m) and 15 per cent (to 5.5 m) respectively.

The joists of a timber floor are not taken into the crosswalls because of the reduction of fire resistance this would cause. Instead they are fitted into mild steel joist hangers built into the wall. As a result no rigidity at all exists at this joint and lateral stability of the building must be provided by other means. Not only is there no rigidity but the floor cannot be assumed to provide any lateral support to the wall with joist hangers only: metal anchors are required for this purpose, at least 30 × 5 mm in section, fastened to the joists and built into or through the crosswall.

Concrete is used for all floors other than those specified above, because the thickness required for an economic span can provide the mass needed for sound insulation. The material is fire resisting and by careful detailing of the bearings sufficient rigidity can be achieved to provide longitudinal stability. Figure 11.1 shows a typical concrete floor-to-crosswall junction, with insulation against sound transmission provided by a cement and sand screed floating on a glass fibre quilt. An important point to note is that the screed must be totally isolated from all other constructions and therefore the quilt must be turned up at the edge as shown.

11.4 Thermal and sound insulation

Some of the problems and their solutions to these two aspects have already been mentioned in this chapter and are dealt with in connection with the infilling and panels in Ch. 12 of *Advanced Building Construction*, Vol. 1, but there is a particular point in the design requiring careful study and that is the junction of the end panels and the crosswalls. The particular problems to be dealt with are those of sound travelling from one flat to the next by a flanking transmission, and heat loss through a thermal bridge formed by the crosswall being taken through to the outside. Figure 11.2 illustrates these problems and several solutions.

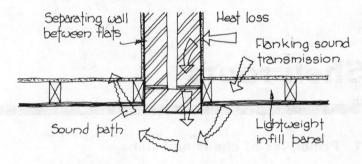

THE PROBLEM

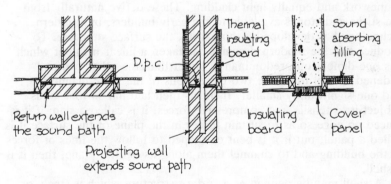

SOLUTIONS

Fig. 11.2   Heat loss and sound transmission in crosswall construction

Chapter 12

Shell roofs

12.1 Principles of shells and slabs

Historically there are two forms of structure, solid – mass construction of stones or blocks where the enclosed space and external shape are not necessarily the same, and skeleton – a light framework and equally light cladding. These derive naturally from the sticks and stones available to the early builders, but modern technology has developed a third form, the surface structure. To produce this the modern designer has taken a linear member which has one dimension predominantly larger than the other two and widened it until there are two dimensions of the same magnitude, and one significantly smaller. If this is then laid flat so that loads subject it to tensile and compressive forces, it is called a slab; if it is placed on edge so as to sustain forces in the plane of the surface it is called a panel; but if it is bent or shaped to follow the lines of forces in the building and to channel them along its neutral plane, then it is a shell.

A shell may be defined as a surface structure which is singly or doubly curved or folded, and the shape is given by the main cross-section. Two main sections can be taken through all shells. In the case of the singly curved shell, only one of these is a curve, and the other is a straight line; in the case of a doubly curved shell both sections are curves; and in the case of a folded shell both are straight line shapes (see Fig. 12.1).

A model of the surface of a single curved or a folded shell can

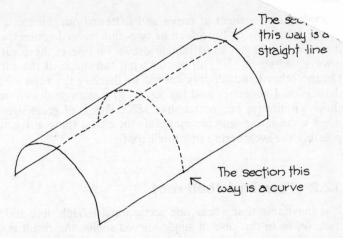

The sec[tion]
this way is a
straight line

The section this
way is a curve

Singly curved shell

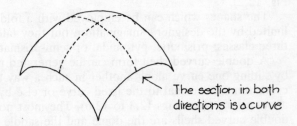

The section in both
directions is a curve

Doubly curved shell

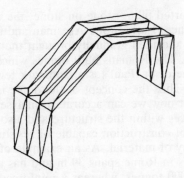

Folded shell

Fig. 12.1 Types of shell

be made from a sheet of paper and flattened out. Hence it conforms to the principles of Euclidean or two-dimensional geometry; for instance, the angles of a triangle drawn on one of these surfaces will always add up to 180°, no matter what the shape of the surface is. On the other hand, doubly curved shells obey the rules of three-dimensional geometry and the angles of a triangle drawn on one of these will always be greater than 180°. This is of great significance when considering suitable materials for either the construction or possibly the weathering of a shell roof.

12.2 Forms of shell roof

It is inevitable that where one section is a straight line and the other a curve, as in the case of single curved shells, the result is some form of barrel vault, the only variation being the nature of the curve (which can be segmented, semicircular, semi-elliptical, parabolic or hyperbolic) or the inclination of the straight line section, to produce a conic shell. Further variations can be produced by adding shells or part shells together or by intersecting two or more shells (see Fig. 12.2).

The shapes which can be employed with a folded surface are only limited by the designer's imagination, but they fall within one of three classes: prismatic, pyramidal or semi-prismatic (see Fig. 12.2).

A double curved shell form can be generated in two ways, either by sliding one curve along another in such a way that the moving curve remains normal to the fixed curve or else by rotating a curve about an axis (see Figs 12.3 to 12.6). The most notable of these double curved shells are the dome and the saddle back or hyperbolic paraboloid (sometimes abreviated to hypar or HP).

12.2.1 Domes

Ever since man started piling stone on stone, the dome has been used to enclose space. Early Stone Age man and the Eskimo built circular structures and then gradually brought the walls over to close the top in a dome. The Romans built domes which still stand today and the great dome of St. Paul's is a landmark few visitors to London will miss. Thus the concept of the dome is not new; all that has changed is that now we can accurately analyse the magnitude and pattern of forces within the structure and also we have at our disposal methods of construction capable of resisting those forces with great economy of material. As an example of this change, the dome of St. Peter's in Rome spans 39 m and has a mass of approximately 10 000 tonnes, whereas a reinforced concrete shell dome of the same span would have a mass of approximately 300 tonnes or one-thirtieth of its sixteenth century counterpart.

The simplest form of dome is the hemisphere formed by rotating

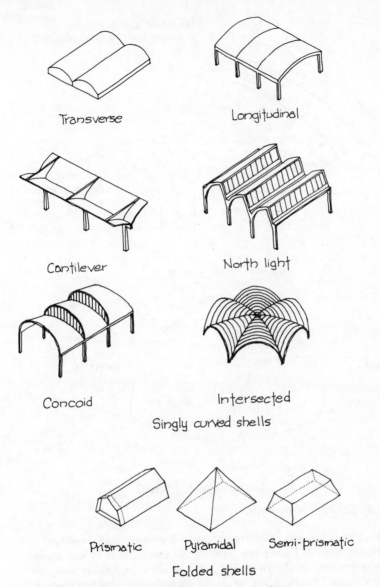

Transverse

Longitudinal

Cantilever

North light

Concoid

Intersected

Singly curved shells

Prismatic

Pyramidal

Semi-prismatic

Folded shells

Fig. 12.2 Forms of shell – singly curved and folded

a semicircle about the vertical axis. Vertical sections in the plane of the axis are semicircles and their perimeters are meridians or lines of longitude on the surface. Horizontal sections are circles and their perimeters are latitudes. It is along these lines that the forces act, as explained in the next section. This form suffers from the very real problem that it requires a circular base on which to sit, and since

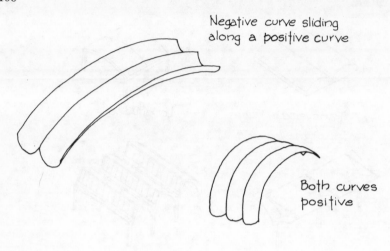

Negative curve sliding
along a positive curve

Both curves
positive

Transitional shells

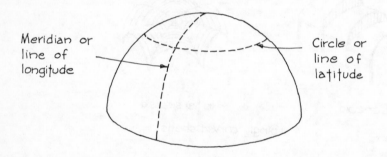

Meridian or
line of
longitude

Circle or
line of
latitude

Rotational shell (dome)

Fig. 12.3 Doubly curved shells

most buildings are square or rectangular a geometrical transition
must be employed. Either the base of the dome must be cut away
until it is square, as shown in Fig. 12.4, or else four spherical
triangles, called pendentives, must be constructed between the
square of the building and the circle of the dome, also shown in
Fig. 12.4.

An alternative dome form is the elliptical paraboloid or
translational dome. This is not a rotational form but is produced by
sliding a plane curve along another of the same shape (see
Fig. 12.1). Being generated by this rectilinear action, this form of
dome naturally assumes a square shape at the base which
conveniently fits a building plan.

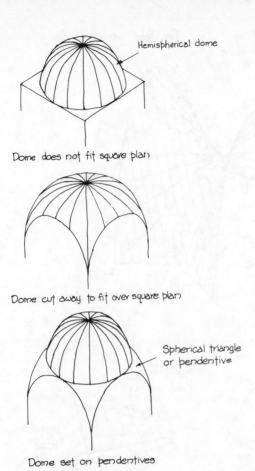

Hemispherical dome

Dome does not fit square plan

Dome cut away to fit over square plan

Spherical triangle
or pendentive

Dome set on pendentives

Fig. 12.4 Forms of dome

12.2.2 Hyperbolic paraboloid shells

A hypar shell (shown in Fig. 12.5 at A) is a shape generated by
moving a negative parabola (ASC) along and at right angles to a
positive parabola (DSB). A positive curve is one which bends in the
direction of the line of force, i.e. downwards, and a negative curve
is one bending upwards. The saddle-back shape so produced is such
that a horizontal section taken through the shell is bounded by a
hyperbola – hence the complicated sounding name – but more
interesting than that is the fact that at every point of the surface
there are two intersecting straight lines which lie in the surface, e.g.
XX and YY. The significance of this fact is that it is possible to
build this interesting, complexly curved and strong shape using
straight members, which has made it a popular roof form.

The complete geometrical figure shown in Fig. 12.5 at A is not
used because of the awkward plan shape required to accommodate

XX and YY are intersecting
straight lines on the surface

Negative parabola

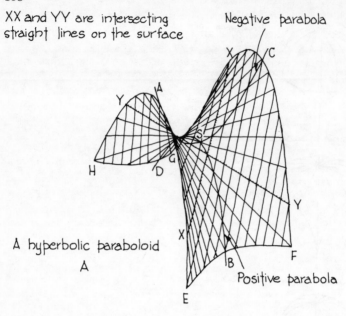

A hyperbolic paraboloid

A

Positive parabola

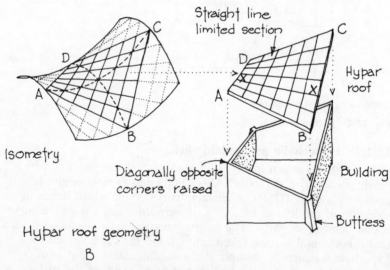

Straight line
limited section

Hypar
roof

Isometry

Diagonally opposite
corners raised

Building

Buttress

Hypar roof geometry

B

Fig. 12.5 The hyperbolic paraboloid shell

the curves EBF and GDH. Instead, a section of the surface is used,
bounded by four of the straight lines which lie on the surface. This
is referred to as a straight line limited section and its development is
shown in Fig. 12.5 at B in which the panel ABCD is cut out of

saddle back shape and placed over the four corners of the building. It will be seen from the diagram that two corners of the building must be higher than the other two to fit the roof shape. The ratio of this rise to the diagonal length determines the rate of curvature of the roof and since, if this falls too low, bending forces will develop in the structure, the lower limit for this ratio is 1:15 for a shell 6 m square, 1:14 for one 12 m square and 1:13 over a building 18 m square.

Instead of using the parabolas of which this shape is composed, it can be readily seen that the straight line limited section can be generated from the straight lines it contains by sliding the line XX along the lines AD and BC in Fig. 12.5B. Or, in more practical terms, if lengths of timber are fixed to the angles shown by the lines AD and BC and narrow boards laid across them, the resulting roof form is a hyperbolic paraboloid. However, there is an inclined outward thrust at the two low corners of this roof necessitating either the buttress shown in Fig. 12.5B or a diagonal tie rod.

Interesting as this building form is, it can be further enhanced for particular plan forms by placing it over a rectangle which is symmetrical about one axis only (see Fig. 12.6A) or by connecting several shells together as shown in Fig. 12.6B, where six shells have produced a roof requiring support at six points only.

12.3 Forces in shells

The calculation of the stress in a shell under a given pattern of loading is a complicated exercise beyond the scope of this chapter, but without indulging in a detailed and advanced mathematical analysis it is possible to appreciate the way that the forces are distributed and the means by which shells are made to carry them.

Curved shells are direct descendants from the stone vaults of ancient times and possess the same stress characteristics. A structure built of stones or bricks laid in mortar is incapable of resisting tensile forces of any appreciable size. Thus a vault built in this way must be designed to be subjected to compressive forces only. Such is the case with a barrel vault in which the imposed load and self weight more or less follow the line of the structure and, as with an arch, force the voussoirs firmly together. In practice, very few vaults were built so that the curve precisely followed the line of stress but, as they were thick, the stress curve was still accommodated within the vault. Modern thin curved shells must be either designed to fit the stress curve or strengthened to cope with additional stress where the line of compressive force leaves the neutral plane of the shell.

The perfect shape for a barrel vault is a catenary curve, which is the line a chain follows when suspended between two points ('catenary' is derived from the Latin word 'catena' meaning 'chain').

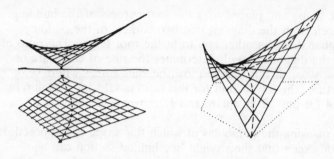

A hyperbolic paraboloid over an asymmetrical plan

A

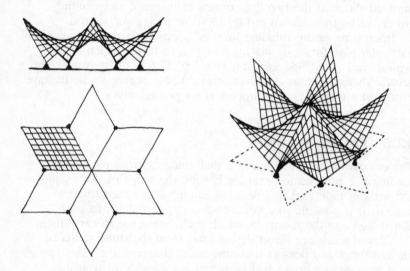

Six connected hyperbolic paraboloid shells

B

Fig. 12.6 Hyperbolic paraboloid roof forms

It follows that if this line, produced by tensile stress acting on an object incapable of resisting compressive forces, is inverted, it is the line to be followed by a member which is incapable of resisting tensile forces but required to carry a compressive stress (see Fig. 12.7).

Two points arise from this statement: whereas the catenary curve is the natural line for a given set of conditions, there is nothing constant about it and if the conditions vary slightly so the curve will

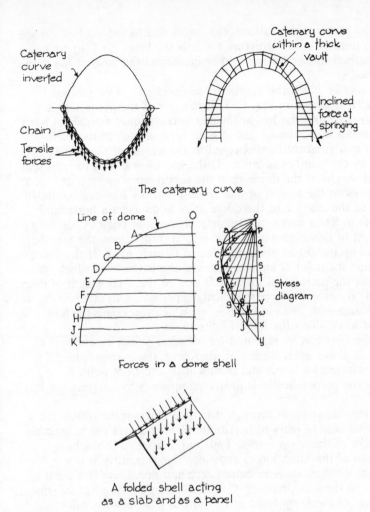

Catenary
curve
inverted

Chain

Tensile
forces

Catenary curve
within a thick
vault

Inclined
force at
springing

The catenary curve

Line of dome

O

A
B
C
D
E
F
G
H
J
K

o
p
q
r
s
t
u
v
w
x
y

b
c
d
e
f
g
h
j

Stress
diagram

Forces in a dome shell

A folded shell acting
as a slab and as a panel

Fig. 12.7 Forces in shells

change. In a building, the loading conditions vary from time to time
but the shape of the roof does not – in most cases! It can also be
seen that when the forces arrive at the springing of the vault or shell
they are not perpendicular and thus tend to overturn the structure
on which the shell sits. Gothic builders constructed great buttresses
to deal with such inclined forces; modern builders can reinforce the
shell to contain the outward movements.

In considering the pattern of forces in a dome it could be
assumed that the same logic used for barrel shells is applicable, but
this ignores the fact that in spreading through a barrel shell the lines
of force remain parallel, while in travelling down a dome they
diverge. The consequence of this is that, even if they are contained

within the plane of the surface, the forces due to the load of a dome induce a tension in the structure towards the base. Sir Christopher Wren realised this and slung iron chains round the base of the dome of St. Paul's.

The way in which this tensile force develops can be plotted graphically as shown in Fig. 12.7. If the line of the dome OK is drawn to scale and the height divided into an equal number of parts by horizontal lines, the weight of each zone or part of the shell between two adjacent lines is equal to the weight of the dome divided by the number of zones. If the line oy is drawn to represent the total weight of the dome then the subdivisions op, pq, qr . . . xy each represent the weight of one zone. From o a line can be drawn parallel to the tangent to the dome at A to intersect a horizontal from p at a. Then oap is the triangle of forces at point A, op being the weight of dome above A, oa the total thrust along the meridian of A and ap the thrust along the line of latitude at A. If the process is repeated for point B another triangle obq is found in which ob represents the meridional force at B and oq the total weight of dome above B. By dropping a perpendicular line from a to intersect bq at b' the distance bb' is found (bq − ap) which represents the hoop force, or force along the line of latitude, at B.

The process can be repeated for all other points except that C and D are above each other and, therefore, the perpendicular line from C intersects a horizontal from s at d, and from point E onwards the perpendicular is projected upwards to intersect the line above.

Working downwards through the force diagram the difference between successive pairs of horizontal lines indicates the magnitude and nature of the hoop forces. Thus, bq − ap gives bb' (the magnitude of the hoop force) and since this is positive it is a compressive stress. ds − cr equals zero and thus there is a point of no stress in the zone between C and D, and gv − fu gives ff' (the magnitude of the hoop force at F) but this is a negative value, indicating a tensile stress. The hoop stress at any line of latitude can be found from the distance between the two curves oy and px drawn through the points a, b, c . . . y and p, b', c' . . . x.

There are two forces in a folded shell. Since any folded shell is composed of a series of flat surfaces these must always act as slabs when subjected to load, i.e. suffering tension at the face furthest from the load and compression at the nearest face. It is true that because of the way the surfaces are disposed in the structure the actual bending forces applied to the slabs are usually small, but they must still be considered. The second set of forces arises through the edge connection of one slab to another and the support the first provides for the second. In this way the slab is acting as a load-bearing panel and thus is subjected to compressive forces along one edge and tensile forces along the other (see Fig. 12.7).

12.4 Materials of construction

The materials from which shell roofs are now constructed are reinforced concrete and timber. The latter tends to be restricted to single curvature shells because doubly curved forms would require the use of tapered boards. An exception to this is the doubly curved hypar shell which, because the surface contains straight lines, can be covered with parallel edged boards.

Concrete shells are extremely thin in comparison to other outer concrete structures. A thickness of 75 mm is quite normal, reducing to 50 mm in parts of the world where the climate is dry and there is less risk of reinforcement corrosion. Unfortunately formwork costs tend to be high, especially with double curved roofs, where 60 to 70 per cent of the total cost is in the formwork and scaffolding supports.

Another problem encountered is that of placing and vibrating the concrete. These procedures must both be carefully carried out, since the very thin section leaves little room for error. Because of the curvature, the concrete consistency should be that of damp earth: if it is wetter than this it will flow down the curve, whereas if it is drier it cannot be satisfactorily worked round the reinforcement.

Thorough vibration is important to ensure a good bond between concrete and steel, but the normal poker vibrators cannot be used because of the thin section and the risk of spattering the mix. Plate vibrators applied to the surface have proved satisfactory, as has a fork shaped vibrator applied to the reinforcement. Where the support structure is adequate, consolidation of the concrete has been achieved by vibrating the scaffolding holding the formwork.

One solution to the problem of concrete placing is to use atmospheric concrete (Gunite). This is a process in which a fine dry mix is blown along a tube by compressed air, mixed with water in a nozzle on the end of the tube, and fired at the reinforcement and formwork. The process is particularly suitable for thin sections and produces a dense concrete with a high degree of resistance to moisture penetration and very little shrinkage.

Several shell roofs, the famous Sydney Opera House shells for example, have been successfully constructed from prefabricated panels held together by post tensioned cables. The saving in formwork costs by this method is considerable, but erection can be difficult and slow.

Chapter 13

Roof lighting and ventilating systems

13.1 Where roof lighting and ventilation are required

The perforation of any roof covering to obtain natural light or
ventilation is a potential source of trouble, since it puts a hole in the
most exposed element of the building. It should, therefore, be
carefully considered whether the roof glazing is required and if so,
how the weather is to be excluded. In many building designs a
balance must be struck between the energy saved in artificial lighting
by the provision of glazing and the energy lost by heat passing out
through that glazing. This is particularly true of roof glazing which,
because of its location, can give rise to a large heat loss.

Bearing all this in mind, the provision of roof lighting is
restricted to single-storey buildings with a large covered area so that
the majority of the floor space is remote from the windows or to
areas of buildings, such as internal corridors, which without roof
glazing would have no natural light at all. In the case of the large
plan, of which a factory production area is a good example, a
further consideration is that the roof lighting can probably only be
considered as general illumination and artificial lighting will still be
required at the work places, mainly because, in the interests of
energy conservation, the area of roof glazing is now restricted to a
maximum of 20 per cent of the roof area by the Building
Regulations.

13.2 Design of roof glazing

Part FF of the first amendment to the Building Regulations
introduced control over the extent of window and roof glazing in
buildings other than dwellings by setting a maximum on the area of

each expressed as a percentage of the roof and wall area. These percentages, which refer to single glazing, are shown below:

	Institutional buildings Other residential Buildings (%)	Offices; Shops; Places of assembly (%)	Factories; Store buildings (%)
Windows	25%	35%	15%
Rooflights	20%	20%	20%

Note that all buildings have a common 20 per cent limit on roof glazing and a variable limit on windows.

These percentages are still applied now in Part L and are used in association with U values for the walls, floor and roof of 0.45 W/m^{2}K to give a 'Notional Building'. All these can be varied in a proposed building but the result must be that either the rate of heat loss or the annual energy consumption would be no greater than for a corresponding Notional Building. Two calculation procedures are given in Approved Document L1. Procedure 1 compares heat loss and Procedure 2 is a similar comparison but based on annual energy consumption.

Procedure 1 would allow an increase in the area of roof glazing, if required, particularly if it is double glazed, by reducing the size of the windows. This may be advantageous in a deep plan building where perimeter daylighting is of very little effect.

As an example consider a factory block measuring 100m wide by 150m long and 5m high in which, say, 60% of the roof was to be covered with double glazed rooflights and the single glazed windows were not required to be any greater than 5% of the wall area.

Table 13.1 below shows the areas of elements in both the proposed building and a Notional Building of the same size, the appropriate U values and the comparative rates of heat loss:

Notional building				Proposed building			
Element	Area [m^2]	U value [W/mK]	Rate of heat loss [W/K]	Element	Area [m^2]	U value [W/mK]	Rate of heat loss [W/K]
Walls	2125	0.45	956	Walls	2375	0.45	1069
Windows (15%)	375	5.70	2138	Windows (5%)	125	5.70	713
Roof	12000	0.45	5400	Roof	6000	0.45	2700
Roof lights (20%)	3000	5.7	17100	Roof lights (60%)	9000	2.00	18000
Floor	15000	0.45	6750	Floor	15000	0.45	6750
		Total	32344			Total	29232

Table 13.1 Comparative rates of heat loss

As can be seen, the total heat loss from the proposed building, even with its very large area of roof lighting, is less than the corresponding Notional Building and, therefore, it meets the Building Regulation requirements.

13.3 Types of glazing in large span roofs

The type of glazing used is determined by the form of the roof, which is dictated by the space to be covered. The simplest roof form has an equal pitch each side of the ridge, referred to as a shed roof. Making the pitches asymmetrical forms a northlight roof and flattening it with raised portions for glazing creates a monitor roof. Typical details are shown in Fig. 13.1.

13.3.1 Shed roof

A steel or aluminium framed roof can span as much as 60 m with a single frame but even when the pitch is kept as low as the roof covering will permit, there is still a large volume of air in the roof space which is absorbing heat from the occupied areas below. Consequently this maximum span is seldom approached, especially as other systems enclose a much smaller roof void. It is, however, a roof form which lends itself to the simple insertion of roof glazing, by substituting translucent roofing sheets for the normal profiled sheets with which the roof is covered. To ensure an even spread of light at the working plane the ratio of roof light spacing to height must be as shown in Fig. 13.2 at A.

13.3.2 North light roof

This is a very common form of industrial roof. It is easy to transport and erect and can be arranged with a lattice girder between the apex and the bottom tie member, so as to cover large areas with very few supports. It also provides large areas of glazing which, unfortunately, in many cases, exceed the maximum now permitted by the new Building Regulation restrictions set out in section 13.2.

The pitch at one side of the ridge is between 20° and 30° and suitable for corrugated roof sheeting and the pitch on the other, north side of the ridge is between 50° and 90°, which means it can be covered with patent glazing in such a way that rain will not penetrate. The steeper pitch also helps to keep the glass clean by causing a fast rainwater run-off, but reduces the coefficient of utilisation of the glazing, making it less efficient for lighting. Figure 13.2 B shows the ratio of roof light spacing to height necessary to achieve an even spread of illumination.

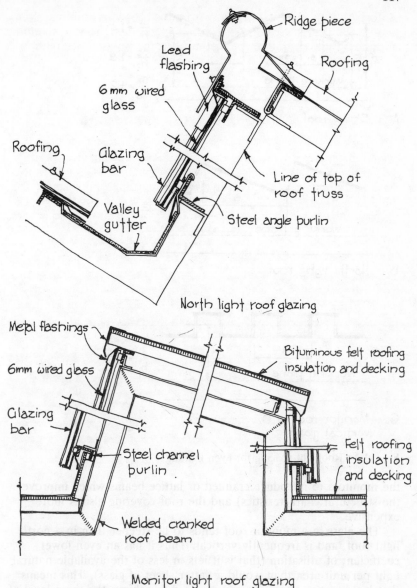

North light roof glazing

Monitor light roof glazing

Fig. 13.1 Typical details of roof glazing

13.3.3 Monitor roof

As this system is essentially a flat roof set with continuous lines of
roof lights it has the advantage over the other two of less roof
volume to heat and a much more presentable soffit. Structurally it is
less economical (although advantage can be taken of the height of

118

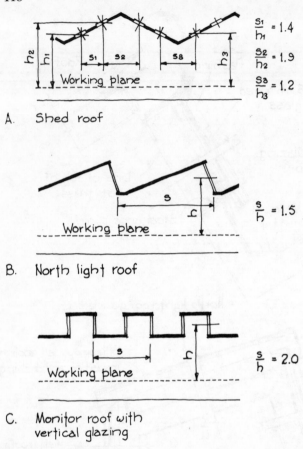

$$\frac{s_1}{h_1} = 1.4$$

$$\frac{s_2}{h_2} = 1.9$$

$$\frac{s_3}{h_3} = 1.2$$

A. Shed roof

$$\frac{s}{h} = 1.5$$

B. North light roof

$$\frac{s}{h} = 2.0$$

C. Monitor roof with vertical glazing

Fig. 13.2 Rooflight spacing for even lighting

the monitors to introduce cranked or lattice beams which improve the structural characteristics) and the roof covering also is more expensive.

The glass in a monitor roof tends to be steeper than in a north light roof, and is frequently vertical. Thus it has an even lower coefficient of utlisation (that is it lets in less of the available natural light per unit area than would a flatter sheet of glass). This means that larger areas of glazing are required to achieve the same level of illumination and can present difficulties in keeping below the Regulation 20 per cent.

An advantage of the monitor roof is that, there being flat areas between the monitors, it is much easier to gain access to clean them and thus maintenance is likely to be more regular and the light admission level kept near the maximum.

Monitors with sloping glass should be spaced at the same ratio to their height as shown in Fig. 13.2, B for north light roofs. When the glass is vertical, the ratio shown in Fig. 13.2, C is required to achieve even illumination.

13.4 Flat roof glazing

The principle of flat roof glazing is to form a hole in the flat roof and then seal it with a transparent or translucent unit in such a way that neither the weather nor burglars can get in. There are many ways in which this can be achieved. Figure 13.3 shows a few, but basically they all consist of a kerb up which the roofing material is dressed and on top of which is fixed the roof light or lantern light.

The kerb may be formed in the roof as shown in Fig. 13.3, by the builder, either with a timber upstand or with one formed in concrete, as is appropriate to the roof construction. Alternatively, in many cases the design of the roof light is based on a metal or glass reinforced plastic kerb intended to sit directly on the roof deck or on the builder's kerb. To conserve heat the purpose-made kerb may be formed hollow and filled with insulating material, and to provide ventilation it may also be fitted with fixed or adjustable louvres or hit-and-miss ventilators.

The kerb may support either a roof light – a single unit of glass or plastic, or a lantern light – an element constructed with glazing bars and wired glass. The former are the more common and generally less troublesome. They are produced in a wide range of profiles – (circular and square segmented domes, hemispherical domes, square or rectangular pyramids, rectangular trapezoid, barrel shaped or flat) and can be installed either as isolated lights or, particularly the barrel shape, in continuous runs. They are also available double skinned to reduce heat loss.

Lantern lights are the older form of roof glazing and usually incorporate vertical glazed sides, which may be openable for ventilation and a pitched, usually hipped, glazed roof. Double glazing is not so readily achieved as in roof lights and consequently there is a high heat loss and condensation can cause discomfort to occupants and disfigurement to paintwork and flooring.

Roof glazing must be studied with respect to security. Roof lights can be unscrewed and glass can be taken out of lantern lights, in each case affording access to an agile burglar. Not much can be done to prevent the removal of glass in lanterns but roof lights can be fixed with specially designed and capped security bolts or with bolts which are concealed from the outside. In both cases a further security measure is to fit burglar bars across the inside of the kerbs, spaced close enough to prevent anybody dropping through.

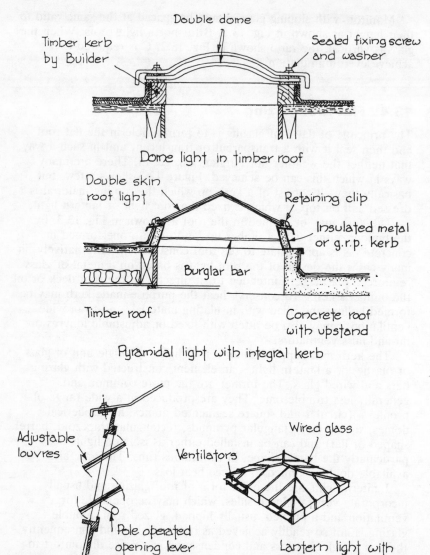

Double dome

Timber kerb by Builder

Sealed fixing screw and washer

Dome light in timber roof

Double skin roof light

Retaining clip

Insulated metal or g.r.p. kerb

Burglar bar

Timber roof

Concrete roof with upstand

Pyramidal light with integral kerb

Adjustable louvres

Wired glass

Ventilators

Pole operated opening lever

Ventilating kerb

Lantern light with glazed and ventilating side walls

Fig. 13.3 Flat-roof glazing

Chapter 14

Roof finishes

14.1 Weather exclusion

The material with which a roof is covered has one main purpose: the exclusion of all types of weather in the severest form likely to be encountered. Thus the roofing must withstand not just the conditions which occur in normal circumstances but the freak weather which is experienced only once every 5 to 10 years. It is of no value to line a roof with a material which will last 60 years, only to have it ripped off the first time the wind gusts to an unusual 50 m/s.

Penetration of a roof covering arises due to either bad design, which allows a combination of rain and wind or snow and wind to find a way through, or a final breakdown following a long period of slow deterioration.

In the first group of roof failures one would class the example cited above of roofing being ripped off. This is a failing to which lightweight coverings and low-pitched roofs are particularly prone. Instances have been recorded of complete aluminium roofs being lifted off buildings and carried some distance in strong winds. With unit roof coverings such as tiles and slates, the efficiency of weather exclusion is related to the pitch, which is determined by the size of the roofing unit, its texture and its accuracy. This is further explained later in this chapter but, in the main, the flatter the pitch the more economical the roofing but the greater the tendency for the slates or tiles to lift in a wind and rain or snow to blow under them.

Gradual deterioration is brought about in all materials by the weather generally, but which particular type of weather is the cause depends on the nature of the roofing material. Sunlight has an adverse effect on anything containing oils and, therefore, leads to a breakdown of bituminous roofings. The sun's heat causes significant thermal movement in metals and plastics, leading to fatigue and eventual rupture. Wind can carry with it dust particles which will erode materials subject to severe exposure. Rain and frost in combination can bring about the delamination of slates and tiles. Salts brought down from a polluted atmosphere by the rain change to weak acids and gradually corrode or dissolve a range of roof coverings.

Add to these the disturbance which can occur through condensation on the underside of flat roofings causing blisters, plants such as ivy growing over and through pitched roofs, premature decay of tile battens, chemical attack of bird droppings and accidental damage by maintenance men and it can be seen that the choice and detailing of a roof covering requires great care.

14.2 Rainwater run-off

To provide complete protection to the building it must be arranged not only that the roofing deflects the rain from the interior but that deflected rain is safely collected and harmlessly disposed of. The quantity of rainwater to be handled is the product of the design of the roof and the intensity of the rainstorm. As already mentioned, roofs (and in this instance gutters and downpipes as well) must withstand the most severe conditions to be anticipated. In England, maximum rainfall is taken to be 75 mm per hour for a duration of 5 minutes once in every 4 years and for 20 minutes once in every 10 years.

Since this occurs during storms (i.e. in association with strong winds) the rain is driven against one of the slopes, if the roof is pitched, further increasing the intensity. The angle of this driving rain is taken to be 60° and to take account of this the actual plan area of one slope is increased to an effective area by an amount equal to half the elevational area (see Fig. 14.1). Of course, the opposite roof slope has an effective area less than the plan area but the worst case must always be used for calculation.

The first step in calculating gutter and downpipe sizes is to convert the rainfall intensity per hour to a quantity per minute per unit area of roof. The extreme case of 75 mm/h is equivalent to 1.25 litre/min falling on a square metre of roof. It simplifies the calculations if a section of a building corresponding to 1 m of eaves length is considered. Taking a building, say, 10 m long × 5 m wide with a roof rise of 2 m from eaves to ridge, the effective roof area per metre of eaves would be 3.5 m² (see Fig. 14.1). If rain falls at the rate given

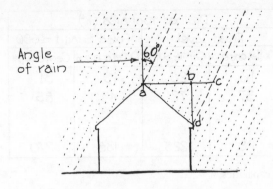

ab = plan width
bd = elevational height
bc = $\frac{bd}{2}$ (1 : 2 : $\sqrt{3}$ triangle bcd)
ac = effective width = ab + $\frac{bd}{2}$
effective area per metre of eaves =
ac × 1.0

Fig. 14.1 Effective roof area

above, the run-off into the gutter on this roof would be 3.5 × 1.25 = 4.375 litre/min or 0.073 litre/s. This is the flow which will occur at the lower end of the first metre of gutter; at the lower end of the second metre the flow will be 0.073 × 2 = 0.146 litre/s and so on. Thus the distance between the end of the gutter and the outlet is determined by the capacity of the gutter to contain the cumulative flow. From the table in Fig. 14.2 it can be seen that a 100 mm gutter has a flow capacity of 0.772 litre/s. If the flow rate per metre is 0.073 litre/s, then $\frac{0.772}{0.073}$ m is the longest the gutter can be without overflowing, i.e. 10.5 m. From this it can be seen that the building being considered requires a 100 mm gutter fixed level to each eave with an outlet at one end.

A simpler method of finding either the required gutter size or outlet spacing is by reference to a table of maximum roof areas as shown in Fig. 14.2. This table gives figures for gutters fixed level and to fall. By inclining the gutter the rainwater is induced to run faster, giving a greater flow capacity and hence a larger drained area of roof.

The figures quoted refer to unplasticised PVC gutters. Other materials possess different flow capacities – for instance the inside of a cast iron gutter is much more rough than that of a UPVC gutter and so the rainwater flows more slowly and the flow capacity is less.

Gutter size	Flow	Drained roof area (m²)			
Nominal	capacity	Fixed level		Fixed to fall 1 in 600	
half round	litres/s	Outlet one end	Centre outlet	Outlet one end	Centre outlet
75	0.333	16	26	19	31
100	0.772	37	72	44	86
True half round					
150	2.385	113	225	135	270

Fig. 14.2 Table of gutter capacities

14.3 Choice of roofing

Apart from the ubiquitous consideration of cost, there are two factors which influence the choice of roof covering: the pitch of the roof and the aesthetic value of the material. The pitch of the roof may be set by a number of factors, and one of these can be the roof covering. Thus for reasons of architectural character a pitched roof may be selected, for aesthetic reasons plain clay tiles could be chosen to cover it, and, for practical reasons of weather exclusion by tiled roofs, a minimum pitch of 40° would be required.

Flat roof coverings do not pose any aesthetic problems and hence the choice rests on cost and suitability. Cost considerations should not be confined to initial cost: cost-in-use should also be taken into account. None of the flat roofing materials require periodic maintenance but they all have a finite life. If this life is less than the projected life of the building then replacement costs must be allowed for when comparing such shorter-lived coverings to more expensive coverings which are more durable. Suitability dictates that the covering is either jointless or else that the joints are sealed or formed clear above the weatherproof surface. It is also necessary to consider the possibility of wear. Roofs to which access is only gained for maintenance purposes do not require to be covered with a hard wearing surface as would a roof on which it is intended people should walk regularly, or sit in chairs. In many cases the materials cannot withstand this latter use and a protective layer of tiles or pavings is needed to prevent the chair legs puncturing the roof covering.

14.4 Roof pitch

Flat roofs are not built dead level but are laid to fall in the direction in which it is desired the rainwater should flow. The rate of fall depends on the trueness of the surface, since any undulations can cancel out the fall and cause ponds to form. A thin roofing material such as bituminous felt or metal will assume the shape of the deck below and if there is a possibility that this may move – as is the case with timber roofs – the fall must be sufficient to overcome any undulations. With a thicker material such as asphalt the problem is not so acute and the fall can be less, as can be seen in the table in Fig. 14.3. Felt and metal coverings on concrete decks can be laid to the same fall as asphalt. In practice the rate of fall is kept to the minimum which will control the flow of water, to avoid the finished weather surface being a long way above the structual deck at the top of the fall. A big difference between deck and weather surface levels can be expensive in materials and it can significantly increase the imposed load if the fall is achieved by a tapered screed. Furthermore it can present problems in the design of the roof verges.

The slope of a pitched roof is mainly controlled by the ability of the joints between the roof covering units to resist driven rain and

Roof covering	Angle: degrees	Pitch: minimum	Rise per metre mm
Asphalt and lead	¾	–	12
Copper and zinc (with drips)	9/10	–	15
Built-up felt	1	–	17
Copper (welted ends)	6	–	110
Corr. asb. cem. (sealed end laps)	10	1 in 11	180
Zinc (welted ends)	14	1 in 8	250
Interlocking conc. tiles	17½	1 in 6	333
Corr. asb. cem.	22½	1 in 5	400
Slates:			
300 – 350mm wide	30	1 in 3.5	588
200 – 250 mm wide	35	1 in 2.8	715
175 mm wide	40	1 in 2.4	830
150 mm wide	45	1 in 2	1000
Single lap tiles	35	1 in 2.8	715
Plain tiles:			
Concrete	35	1 in 2.8	715
Clay	40	1 in 2.4	830

Fig. 14.3 Table of roof slopes

snow. In slightly porous materials, such as tiles, a secondary consideration is the need to remove the rainwater with sufficient speed to prevent it soaking in, which, in frosty weather, can lead to delamination.

In the case of profiled roofing sheets such as corrugated steel or asbestos cement, sealing of the overlap between the ends of the sheet allows the pitch to be as low as 1:11 or 10° as shown in Fig. 14.3, whereas without this seal the minimum pitch is 1:5 or $22\frac{1}{2}°$.

Single lap interlocking tiles, which are now produced in concrete, can be laid to a pitch as low as 1:15 or $17\frac{1}{2}°$ because the design of the tile provides a profiled section and side lock, which prevents rain water flowing over the edges, and an accurately fitting head lock where the tiles overlap, which prevents driven rain penetrating. Single lap tiles without any interlocking grooves, such as the old type of pantile, cannot be laid to less than 1:2.8 or 35°. Single lap tile profiles are shown in Fig. 14.4.

Slates and plain tiling are laid to double lap and the minimum pitch is set by the closeness of fit between head and tail, the flatness of the surface and the size of the unit. The closeness of fit determines the extent to which rain and snow can be driven up the overlap (in which respect slates are superior to tiles) and the flatter the surface the nearer it is to the pitch of the rafters. In all cases of overlapped roofing units the pitch of the unit is flatter than the pitch of the roof because the head of one unit lifts the tail of the unit above. Added to this, if the surface is cambered – as it is on a tile – the curve steepens the pitch of the unit at the lower end but flattens it still further at the head end. For these reasons, slates can accept a lower pitch than tiles.

The size of slates and tiles affects the pitch because of the way rainwater disperses below the tail of the slate or tile. This is shown in Fig. 14.5 and arises when rainwater runs off the tail of one slate or tile, enters the open joint between the sides of the slates or tiles below and spreads out on the ones below them. What must be achieved is an angle of dispersal of the rainwater which causes it to meet the edge of the slate or tile below the line of the heads of the slates or tiles below. The pitch of the roof affects this angle, in that the steeper the roof the narrower the angle, and consequently the length and width of the slate or tile, and the pitch of the roof are interrelated.

14.5 Roofing materials

References have been made to various materials in the proceeding sections of this chapter. Below is a summary of the characteristics of the materials most commonly used for roof covering.

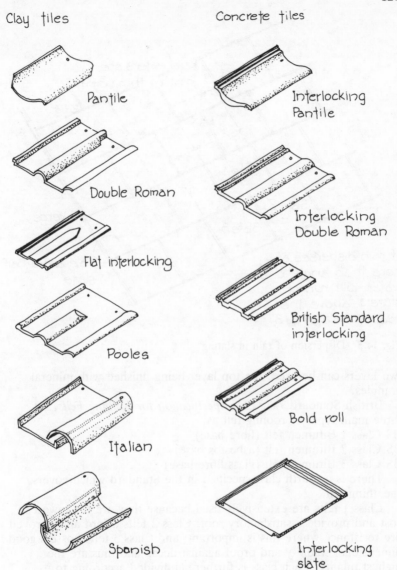

Clay tiles

Pantile

Double Roman

Flat interlocking

Pooles

Italian

Spanish

Concrete tiles

Interlocking
Pantile

Interlocking
Double Roman

British Standard
interlocking

Bold roll

Interlocking
slate

Fig. 14.4 Single-lap tiles

14.5.1 Built-up felt roofing

This is very widely used for flat roofs because it is the cheapest
material for this purpose, but it has a life limited to 15 to 20 years.
When used on flat roofs the covering is built up from three layers of
bituminous felt bonded together with hot bitumen. For pitched roofs

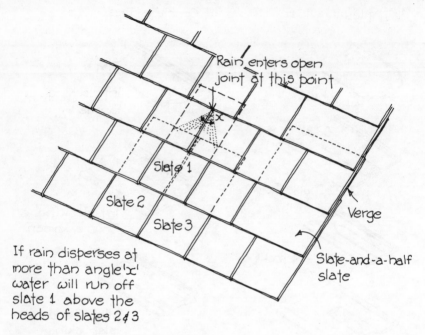

Rain enters open joint at this point

x

Slate 1

Slate 2

Slate 3

Verge

Slate-and-a-half slate

If rain disperses at more than angle 'x' water will run off slate 1 above the heads of slates 2 & 3

Fig. 14.5 Dispersion of rain in slating

two layers can be used, the top layer being finished with mineral granules.

British Standard 747: 1977 *Specification for Roofing Felts*, list three main types of roofing felt as:

BS Class 1 Bitumen felt (fibre base)
BS Class 2 Bitumen felt (asbestos base)
BS Class 3 Bitumen felt (glass fibre base)

There is a fourth class specified in the Standard which covers sheathing felt.

Class 1 felts are extensively used because they are the lowest in cost and provide a satisfactory roof; Class 2 felts afford an improved fire resistance where this is important; and Class 3 felts possess good dimensional stability and proof against decay and thus are the highest quality. Each class is further subdivided according to its finish and its nominal weight per 10 m² (which is approximately equivalent to an 11 m roll 0.9 m wide). The variation in weight is due either to difference in the thickness of the felt (14 to 25 kg/10 m²) or to the mineral granule surfacing (28–38 kg/10 m²). These can all be used in varying combinations and thus a wide range of specifications is possible but in all cases, unless the top layer is mineral finished, a covering of spar chippings is required to shield the felt against the harmful ultra violet light of the sun.

14.5.2 Single layer sheet roofing

A number of proprietary products exist which are similar in their application to built-up roofing but achieve their objective with a single layer. One such material is marketed under the trade name of Nuralite and is a bitumen and mineral fibre roofing of laminar construction. Unlike the felts it is relatively rigid when cold but softens easily with a blow-lamp. Joints are formed by interleaving a welding plastic and applying a heated iron or blow-lamp.

Marleydek, manufactured by Marley, is one of a range of PVC sheet roofings. In this case it is combined with asbestos though in other makes the vinyl is either used on its own or in conjunction with pitch, or with natural or synthetic fibres. Joints are solvent welded and the finished covering possesses high puncture resistance, elasticity and flexibility.

There are also a range of plastic and synthetic rubber or elastomeric roofings used either alone or in combination with asbestos or synthetic fibres. Generally their performance is equal to the roof coverings described above, but their cost tends to be higher.

14.5.3 Sheet metal roofing

The metals used for this purpose are lead, copper, zinc and aluminium. Although lead is extremely durable and can be expected to last 100 years it is very expensive and very heavy, and consequently its use is limited to flashings where its high degree of malleability makes complex shapes possible without jointing. Thickness is defined by a code number which corresponds to its former measure in pounds per square foot (thus code 4 lead weighs 4 lb/ft^2). The recommended thickness for flat roofs is between code 4 and code 7 depending on the roof size and possibility of foot traffic over it.

Copper is quite widely adopted when a metal roofing is specified. In use, the colour at first gradually darkens and blackens, due to sulphurous gases in the atmosphere and then proceeds to turn green. This copper salt surface is permanent and insoluble and prevents any further chemical attack. The Copper Development Association recommend roofing sheet thicknesses of 0.45, 0.60 and 0.70 mm and sheet sizes of 1800 mm long × 600, 600 and 750 mm respectively. Fixing should be by copper clips folded into the seams between sheets and the roof covering should be laid over a class 4 felt (type 4A (II) brown No. 1 inodorous) to reduce possible abrasion between the copper and the deck and to prevent galvanic action between the copper and any steel nails or screws used to fix the deck.

Zinc is used for roofing either in the state it leaves the smelting works, or as a zinc/copper/titanium alloy. It used to be specified by the English Zinc Gauge, which was peculiar to zinc and did not correspond to any of the other metal or wire gauges. Now it is specified by its thickness in millimetres, and should not be less than 1 mm for roofing purposes. The method of use is very similar to copper and it is usually less expensive. However, it is readily attacked by acids and various soluble salts, and easily forms an electrolytic bond with other metals particularly copper, causing rapid corrosion. Great care must, therefore, be taken to avoid any contact with oak or western red cedar, since these are acidic timbers; with materials containing soluble salts, chlorides and sulphates; with gypsum plaster; and with copper and copper-rich alloys. Although it has a lower initial cost than other metals its shorter life expectancy of 40 years can cancel this advantage if the overall costs-in-use are considered.

Aluminium was originally so expensive that its use for roofing was never considered, but improved and increased production have reduced costs to a competitive level which, when linked to the long life of the material and its low weight (which saves on installation and cost of support structure) represents an attractive alternative to the more traditional metals. It is available as 'super-purity' (99 per cent pure) or 'alloy'. The super purity is very malleable and employed for flashings but it is not as strong as the alloy, which is used for roofing. The recommended thickness is 0.7 or 0.9 mm and the methods of fixing and joining are similar to copper. As with zinc, acidic timbers will attack aluminium and contact with copper or water which has flowed over copper sheeting or through copper pipes will cause corrosion.

As can be seen from Fig. 14.3, all metal roofing materials can be laid to a lower pitch with drips than with welted ends. A welt is formed by placing the turned-up ends of adjoining sheets together, folding the two thicknesses over twice to form a double welt and dressing it flat. With lower pitched roofs there is a danger of rainwater being held against the welt and finding its way through the folds. In these situations drips are formed by creating a step in the roof surface up which the lower sheet is dressed. At the top of the step the lower and upper sheets are connected by a single welt formed as described above but folded over only once.

Side joints between sheets are made either by a standing seam, which is like a welt but not dressed down, or, where there is a possibility that this may be trodden flat, by a wood roll. Welts, a standing seam and a batten roll are all shown in Fig. 14.6.

14.5.4 Corrugated sheet roofing

The principle of corrugating sheets of a thin material as a roof covering is the same as that for folded structures discussed in Chapter 12. By taking a thin flexible sheet and folding it up and down into ridges and troughs a component is formed which is sufficiently stiff in its length to carry roof loads across an economic purlin spacing. This is achieved by the near vertical part of each fold

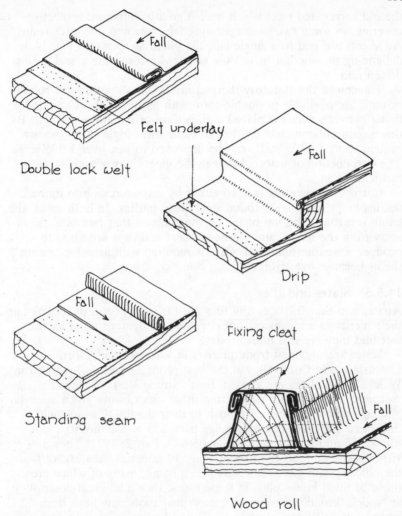

Fall

Felt underlay

Double lock welt

Fall

Drip

Fall

Fixing cleat

Standing seam

Fall

Wood roll

Fig. 14.6 Metal roofing details

or corrugation acting as a narrow beam which is stiffened laterally by
the horizontal part of the corrugations at the top and the bottom.

The oldest material to which this principle was applied was steel,
which is still produced but, unlike its predecessor, is now galvanised.
Following steel, a paste of asbestos and cement was pressed into
corrugated roofing sheets and, being of low cost, very strong, easy to
work and durable, proved very popular (as thousands of industrial
roofs can testify). Recent anxiety about asbestos has reduced
demand. When aluminium became cheap it was used for this
purpose and is rapidly gaining favour, as are the modern versions of

the old corrugated steel which have a greatly enhanced protective covering, in some cases incorporating felt, zinc and polyester resin. All sheets are laid to a single lap, the pitch, as shown in Fig. 14.3, depending on whether or not the end laps incorporate a seal against driven rain.

To achieve the statutory thermal insulation all corrugated roofing systems are available in double form with an upper sheet and under lining between which is placed a glass fibre or mineral wool mat. By this 'sandwich' construction the 'U' value of a single skin asbestos cement roof, 7.95 W/m^2K, can be improved to just over 1.0 W/m^2K. The introduction of spacers between the sheets improves this still further.

Corrugated sheeting is fixed either by drive screws into timber purlins or by hook bolts round steel angle purlins. In both cases, the fixing is through the top of the corrugations so that rainwater flows away from the fixing, and plastic washers and caps are fitted to produce a weathertight seal. Care is required with asbestos cement sheets because overtightening can cause cracking.

14.5.5 Slates and tiles

Apart from thatch, slates and tiles are the oldest form of roofing and their merits as a means of roof covering are demonstrated by the fact that they are still in use today.

Slates are obtained from quarries in Westmorland, North Lancashire and Cornwall, but the best roofing slates are quarried in Wales. Welsh slates are graded 'Best','Strong Best', 'Medium', 'Seconds' and 'Thirds'. Slates from other quarries are given a similar grading, which refers not so much to their durability as to their thickness. The sizes of slates range from 255 × 150 mm to 610 × 355 mm and are set out in BS 680: Part 2: 1971. They are virtually the same as the former range of imperial sizes except that the older system gave each slate size a name, many of which are those of titled ladies plus, in some cases, such additional description as 'wide', 'broad' or 'narrow', etc, so that roofs may have been covered with 'Wide Countesses', 'Broad Ladies', 'Small Duchesses', etc!

Fixing of slates is by nailing, preferably with large headed yellow metal nails (an alloy of 63% copper, 36% zinc and 1% tin), to slate battens, minimum size 38 × 19 mm. They can be head nailed, where the nail hole is made, by the slater, 25 to 38 mm down from the head, or centre nailed. Head nailing provides a double lap to the nail thereby giving this, the weak point in the system, maximum protection. Centre nailing reduces the leverage on the nail caused by the slate lifting in a wind but the nail only has a single slate covering it.

Tiles, originally made from clay and now also produced from concrete, are either single lap (interlocking tiles) or double lap (plain tiles) as already mentioned. Plain tiles are slightly cambered to make sure that the tail sits firmly on the tile below, and are usually provided with two nibs to hook over the tile batten and two nail holes. It is only on severely exposed roof slopes that nailing every course is necessary. Exposed slopes should be nailed every other course, sheltered slopes every third course and very sheltered situations only require nails every fifth course. Eaves courses and verge tiles should all be nailed. Because the tiles in each course are staggered it is necessary to provide a wider tile to produce a straight line finish at the verges. This is a 'tile-and-a-half tile' – the same applies to slating (see Fig. 14.5).

Profiled tiles, for single lap roofing, are produced in a variety of shapes as shown in Fig. 14.4 in both clay and concrete. They are also provided with nibs and nail holes, but some tile profiles have two of each and others have only one.

The nails for fixing tiles should be 38 mm for plain tiling and 64 mm for interlocking and are usually aluminium alloy or 'composition'; aluminium, zinc and copper nails are also satisfactory but galvanised steel nails are not. Battens should be not less than 25 × 19 mm for plain tiles and 38 × 19 mm for interlocking tiles, but wider battens, up to 50 mm, are preferable since they are less likely to split. If the span of the batten is greater than 400 mm between rafter centres the thickness should be increased to 25 mm.

Chapter 15

Curtain walling

15.1 Definition

Figure 12.1 in *Advanced Building Construction*, Vol. 1 illustrated the difference between infill panels, which are fitted between the members of a structural frame, and claddings which are fixed to the face of the structural members or the projected edge of the floor. Curtain walling is a particular form of cladding made up from a grid of metal or timber framing members into which are fitted glass or solid panels to produce a lightweight non-load bearing building envelope, capable of modifying the external conditions to the desired internal environment.

Being non-load-bearing, curtain walling must be supported by, rather than support, the structure of the building and is totally dependent on the structural frame or floors for its strength. Most systems are also incapable of providing any stiffening or lateral restraint to the structure. All curtain walls must be strong enough to convey the stresses due to wind to the points of support and, since the effect of the wind may be a positive pressure or a negative suction, the design of these points of support must be carefully studied.

The vertical members of the grid are usually the strongest because the point of support most commonly used is the edge of the floor and thus the curtain wall is required to span in a vertical direction. These members are called 'mullions' and may be further described as split mullions where, for purposes of installation, they are in two pieces. The horizontal members are referred to as 'transoms' and are usually smaller than the mullions to which they are fixed.

15.2 Advantages of curtain walling

The principal advantages which can be claimed for curtain walling are lightness, thinness, flexibility of design, speed of erection and economy.

Lightness is not particularly significant unless the building is tall. The saving in weight over other systems of building enclosure makes very little impression on the design of the frame or the foundations of a two- or three-storey building, but the greater accumulation of weight saving over, say, 15 or 20 storeys can effect economies in both of these principal elements.

Thinness is not a structural or building advantage but it is a financial one in building developments where the extent of the building is restricted and the rentable value of the floor area is high. In these circumstances, thin walls mean more floor area for the same overall building size, leading to a greater financial return for the building owner.

Flexibility of design arises from the absence of any load bearing elements and thus the design of the curtain walling and fenestration can more easily follow the dictates of the internal plan, the daylighting requirements and aesthetic considerations of the external appearance. Speed of erection, however, is the greatest advantage of curtain walling. The saving in time on site leads to a shorter building programme and earlier occupation of the building.

The economic potential of curtain walling is not in the system itself but in the associated savings described in the preceding paragraphs. Indeed, due to the high production standards which must be maintained for the system to be efficient, the cost of curtain walling can exceed that of more traditional methods of enclosure, and therefore, to gain any real benefit the building design and programme must make maximum use of the other advantages listed.

15.3 Performance requirements

As a lightweight building enclosure, any system of curtain walling must be carefully examined to check that it achieves the performance necessary to produce the desired results. The more general of these are: fixing tolerances and movement control, satisfactory fixing and jointing arrangements, insulation against both heat loss and noise, and resistance to weather and fire.

15.3.1 Tolerances

As already mentioned (see section 4.4) it has been found impossible to match on-site precision to factory production standards and, therefore, tolerances must be allowed in the design of any industrialised components. For this purpose a positional tolerance would be allowed and the members would be designed to function equally well if they were fixed within this positional tolerance on one side or the other of the designed position. Clearly, if the member is moved further off its correct position than allowed, then the fixings will not fit and the members will not be able to be secured. The amount of tolerance depends on the design of the curtain walling, which limits the extent of the acceptable variation, and the type of construction of the building, which sets an achievable standard of accuracy.

Another, much smaller allowance which must be made is a manufacturing tolerance. Even though a characteristic of factory-made products is their consistent accuracy, nothing is perfect and limits must be set to the degree of variation which can be accommodated, bearing in mind that the system must still work even when the worst combination of tolerances occurs.

15.3.2 Movement control

Relative movement must be anticipated between the curtain walling and the building structure because thermal variations produce greater changes in the curtain walling than in the structure, and because the structure will settle to some extent whereas the curtain wall will not. The latter is likely to occur only once but the former is a continuous movement which must be covered in the design and installation.

Not only will relative movement occur between the curtain walling and the structure but also between the parts of the curtain walling system. The panels will move in the framing, and to differing extents. To allow for this, fixing clearances must be left and the fixing beads must be wide enough to bridge both this and any positional variation in the framing member. Clearance for clear glass should be 3 mm for panes up to 760 mm, and 5 mm over 760 mm, off each edge. Any other type of glass should have a 6 mm edge clearance. From this can be found the minimum width for the glazing beads. The maximum arises from the need to restrict the shading of the edge of the glass to a width of 10 mm to prevent a critical build-up of internal stress due to differences in temperature across the pane.

Distortion and bulging of composite solid panels can follow expansion of any core material or entrapped air and differential expansion of the inner and outer surfaces. To prevent this, the two skins should be separate or attached by sliding fixings and the panel should not be hermetically sealed. Shaping the outer skin to a profile will strengthen it and also allow it to take up movement within itself.

15.3.3 Fixing

To provide for the tolerances and movements already described, the fixings must be designed to be adjustable initially and movable when required, but at the same time they must secure the curtain walling to the structure with sufficient strength to give adequate resistance to wind forces. To ensure that the fixings are safe, they should be so designed that only two-thirds of them are needed to withstand the anticipated wind load. This guards against either a fixing failure leading to progressive failure or to damage due to a freak gust of wind.

These requirements are achieved by the use of slot fixings: bolts in slotted holes and packing shims as shown in Fig. 15.1. Where a design relies on the movement of bolts in slots, polytetrafluorethylene (PTFE) washers should be fitted to reduce frictional resistance.

15.3.4 Jointing

The method by which the frame members are connected and the panels fixed in place varies according to the material used and the

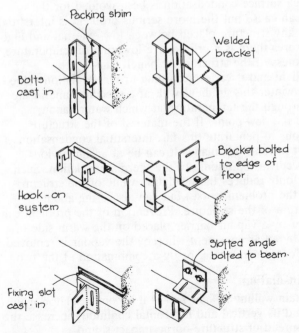

Fig. 15.1 Curtain wall fixing methods

manufacturer. With aluminium framing (one of the most frequently used) the system usually employed is to screw spigot pieces to the sides of the mullions over which the hollow transom is fixed. Similar spigot pieces are set inside the end of one length of mullion to connnect it to the next. Glazing is by non-setting mastic in either a channel section or behind clip-in glazing beads.

15.3.5 Thermal insulation

This is a critical design factor in curtain walling since the basic concept is inclined to be poor in this respect. There are three heat loss routes to consider: through the glass, through the panels and through the framing members. The first is readily dealt with by double glazing and the second by double-skinned panels filled with an insulating material. The third, the so-called 'cold bridge', requires careful design of the section of the framing members if they are of metal. The most efficient solution to the cold bridge problem is a two-part section in which the outer part is connected to the inner part by an insulating core (see Fig. 15.2). An alternative is to apply insulation to the internal faces of normal framing. This, however, makes the curtain wall members much larger and one of the attractive features, its light appearance, is lost.

Without any treatment, the cold bridge will lead to a substantial heat loss and to the formation of condensation on the inner surfaces of the frame. Such surface condensation is best avoided for the disfigurement it can cause but the more serious problem is interstitial condensation. In any structure placed between low external and high internal temperatures there is a progressive lowering of temperature through the thickness of the structure or panel. This is the temperature gradient and it varies according to the insulation value of each layer. In winter this gradient will, at a specific plane in the structure, pass through the temperature at which water vapour condenses, called the dew-point. If the material of the structure allows water vapour to penetrate this far, interstitial condensation occurs at and just beyond this plane. It can be at an impervious layer or it may occur in the middle of the thermal insulation. Such condensation not only reduces the insulation value of the structure (thereby making the problem worse) but can also bring about physical depreciation of the structure. Prevention of the problem can be achieved either by a vapour barrier placed on the warm side of the insulation or by an arrangement whereby the vapour is removed by ventilation before it condenses, or by a combination of the two.

15.3.6 Sound insulation

The fact that curtain walling is light makes it a poor insulator against airborne sound and its vertical and horizontal continuity increase the problems of the spread of structure-borne impact sound.

Discontinuity could solve both problems but it is difficult to

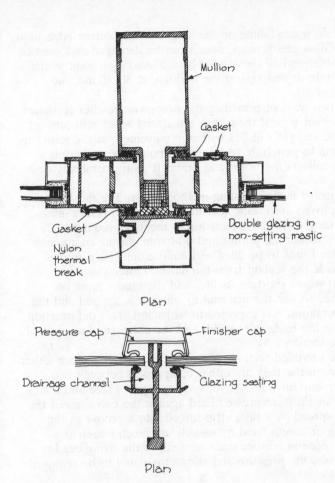

Mullion

Gasket

Gasket

Nylon thermal break

Double glazing in non-setting mastic

Plan

Pressure cap

Finisher cap

Drainage channel

Glazing seating

Plan

Fig. 15.2 Curtain walling details

achieve in curtain walling, whereas the introduction of sound absorbing mass walls behind the curtain will cut down on the airborne sound and may also be required for fire resistance purposes.

15.3.7 Weather resistance

The materials of which curtain walling is made – glass, aluminium, plastics, etc., are all inherently resistant to the effects of weather and therefore the designer must concentrate on the details of the joints (of which there are many) to ensure that they reach a similar standard.

Mention has already been made of the movements to be expected in curtain walling. This movement also affects the joints. The glazing and the panels are smooth and impervious and therefore discharge all rain water falling on them against the bottom edge joint very quickly. Glass gets broken, panels can be damaged and must be easily replaceable, and all these must be achieved by a joint which can withstand rain driven against the building at 30–40 m/s and rising currents of air.

There are two ways of providing the necessary weather resistance at the joints: either to seal them with a material which will prevent any penetration and yet will flex with any movement at the joint, or to allow the rain to penetrate to a selected point where there is a channel which collects the water and drains it away harmlessly (see Fig. 15.2).

A wide range of flexible sealing methods exist. The most commonly employed are mastic sealants or gaskets. Mastics must remain non-setting throughout their life to meet the required flexibility and thus normal putties and oil-based glazing compounds are not suitable. Joints to be filled with mastic must be carefully designed to shade the sealant from the harmful effects of direct ultraviolet light which shortens its life, and the mastic must be carefully selected to suit the material to which it is applied and the degree of deformation it is expected to withstand. This deformation is dependent on the basic joint width and the anticipated amount of movement (see section 9.5).

Gaskets are extruded sections of neoprene or vinyl and are either solid or hollow. In use they are either compressed between two members of the curtain wall system, such as a panel face and a clip-in fixing bead, or they are pressed hard against the two sides of the joint by being spread by a filler strip forced into a groove in the gasket. Glazing or panels fixed by gaskets are much easier to maintain than those in mastics since removal of the fixing bead or filler strip releases the pressure and allows the panel to be removed.

15.3.8 Fire resistance

Curtain walling usually employs unprotected steel, aluminium or timber framing in small sections to support the infilling panels and can in no way meet fire resisting standards on its own, no matter what degree of fire resistance is inherent in the panels. The only satisfactory way to provide fire resistance of a standard required by the Building Regulations is to build a fire resisting back-up wall or panel behind the curtain wall panels, thus reducing the 'unprotected areas' described in the Regulations to just that of the clear glazing.

Part IV

Internal construction

Chapter 16

Partitions

16.1 Partition types and function

Partitions are any internal structures provided to sub-divide the space enclosed by the external building envelope. They may also be required to provide support for other elements, such as floors, roofs, or other partitions above. Within the non-load bearing classifications there are three types: fixed partitions, that is, ones which if moved leave damage behind; demountable or relocatable partitions which can be moved when a re-division of the enclosed space is required; and movable partitions, which easily and quickly can be moved aside and as readily returned to their original position.

All types of non-load bearing partition can be constructed with either metal or timber framing and a wide variety of facings. Fixed partitions can, in addition, be solidly constructed of brick or, more commonly lightweight concrete blocks. The final choice is dictated by the performance standards as set out in section 16.3 and by the type of building in which the partitioning is to be placed. The most logical partition structure is one which is an extension of the type of construction used in the building – if it is a brick structure, block partitions would give similar performance whereas a timber structure, which is slightly flexible, would be better with timber framed partitions. In many cases, where the building is a large framed structure this logic does not apply or may be overtaken by other considerations, such as a demand for plan flexibility which would require relocatable partitions.

This requirement of a flexibility in the plan arrangement must be examined carefully when selecting a partition system. Clearly no form of fixed partitioning would be suitable but when considering the non-fixed partitions, a statement is required of the reason for the flexibility, which would indicate the frequency with which the changes are likely to occur and the ease with which they must be effected. At one extreme of the scale of flexibility is, say, a multi-storey office building erected as a speculative investment for letting. This probably would be left completely free of any internal sub-divisions, except those required for safety, i.e. compartment walls, or for privacy in areas of sanitary accommodation. When let, the tenant would install partitioning in the building to his own requirements, which might well remain in place until he relinquishes his tenancy, at which time it is probable that the succeeding occupant would change it all. This time scale is in years or even decades. At the other extreme is, say, a multi-purpose community hall which may be used as one large hall on one day, divided into two smaller rooms for two functions the next day, and restored to one hall the following day.

16.2 Partitioning methods

In very many cases, the various ways in which a partition can be formed can be used either to construct it in position, or as the basis for a factory-produced, proprietary system of partitioning units. The following is a selection of the many methods available and their application to the three partition types.

16.2.1 Fixed partitions

These can be of brick or block construction, plastered or dry lined. Their choice is favoured where wet trades are already involved with the construction and the partition units would bond with the structure. Lightweight concrete block partitions can be built directly off a suspended floor in many structures, but timber floors need to have double joists on the line of the partition, for additional strength and stiffness, and some concrete floor systems may also need local strengthening. Brick partitions in upper storeys require supporting beams, unless they are placed over a similar partition in the storey below.

Timber and metal framed partitions, built in position and lined with plasterboard or metal lath and plaster, cannot be considered as relocatable because very little re-useable material can be recovered if the partition is taken out.

16.2.2 Demountable partitions

These, like fixed partitions, can be made up with timber or metal studwork, with a facing of a suitable board, the differences between demountable and fixed partitioning in this instance being mainly in the method of manufacture and in the finishing. Demountable partitions are made up in panels off site, possibly purpose built for a particular location, but usually manufactured as a standard range. Since they are intended to be capable of being taken apart, all finishings must stop at the edges of the individual panels (a continuous finish like plaster is not suitable) and the inevitable joints must be treated, sealed or concealed by an arrangement which is removable.

In some partitioning systems the frame is exposed and infilled with solid or glazed panels and in others, where the partition comprises panels of two sheets of plasterboard or other rigid sheet material with a core structure, the studs required are more of the nature of connecting pieces than a support framework (see Fig. 16.1).

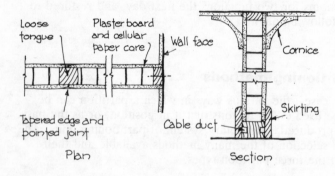

Plasterboard sandwich partition

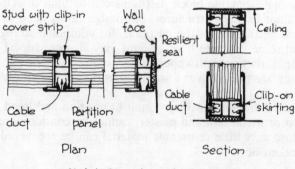

Metal framed partition

Fig. 16.1 Partition details

16.2.3 Movable partitions

The panels of which these partitions are formed have more in common with doors than with walls. The most common form consists of a top and bottom track into which are fitted the partitioning panels, and which slide past each other, or slide and fold. There is a wide choice of panel, either timber or metal framed, finished in natural timber, or lined with decorative boards or vinyl sheet.

The system is quick and easy to operate (if well maintained) but the partition can only be moved along a predetermined line. An alternative method uses similar panels but, instead of being fitted to tracks, the panels incorporate an expanding top edge, usually operated by a key or winding handle inserted in a vertical edge. In use, these panels are placed along the line where the division is required and the handle inserted and turned. This has the effect of extending the top edge until it is pressed hard against the ceiling and the panel is wedged firmly in place. The handle is then removed and the next panel fitted to the first. As these panels are held in place by their tight fit, they cannot be used with a suspended ceiling. Where the room to be divided is high, the size of the panels can make handling difficult and storage a problem, but they do possess the unique advantage of total flexibility and no resultant damage to the adjoining finishes.

16.3 Partition performance standards

As well as sub-dividing the enclosed space, partitions are also required to meet a number of other criteria. Some are peculiar to the particular building, but most partitions are required to achieve a specified standard of finish and must provide means of installing services. In addition, they may also have to possess resistance to fire and sound, and provide a means of lighting or ventilating the sub-divided space. Any openings for glazing or for ventilation purposes must be carefully studied for the effect they may have on the fire and sound resistance values of the partition.

16.3.1 Finishing

The finishes which can be applied depend on the type of partition. A fixed partition can be treated with a continuous finish, such as plaster, which can then be decorated with any of the many materials on the market for this purpose. Demountable partition panels, of timber or metal studding, are usually lined with plasterboard or a similar rigid sheet material, faced with vinyl, hessian or a variety of veneers before they leave the manufacturer. In this case, the finish must be capable of withstanding the effects of both the normal daily use of the building and the process of installation, disassembly and re-erection. Movable partitions are frequently handled and, therefore, require a finish which will not readily show finger marks.

Within these constraints there is a wide choice of material, colour and texture which is solely dictated by the likes and dislikes of the building owner.

16.3.2 Accommodation of services

Fixed partitions can have built into them service ducts in which both pipes and cables can be run. Demountable partitions are rendered ineffective as relocatable divisions if pipes are fitted into, or on to them, but cables for electricity or telephone purposes can easily be withdrawn, allowing the partitioning to be changed. Movable partitions cannot accommodate any services.

The most common method used for running electrical services in demountable partitions is to arrange wiring ducts in the cill member and the vertical framing members. These ducts are closed by clip-on skirtings and vertical cover strips (see Fig. 16.1). By this means, the wiring can be installed to all the points specified at the time of erection and furthermore, since the skirting and cover strips can be unclipped, the services can be altered or extended with greater ease than with fixed partitions.

16.3.3 Fire resistance

In most cases the fire resisting qualities of a partition are of importance and practically all systems will achieve at least a half hour standard of resistance. An exception to this is a glazed partition with plain glass. Glazing with wired glass can give a half hour resistance and if the glazing beads are of metal, capped with metal or painted with intumescent paint, the period is increased to one hour. In addition, the maximum area of each pane must not exceed 1.1 m^2, and the lowest level of glazing should be 1.2 m above the floor, if the partition lines an escape route, to allow occupants to crawl below the radiated heat of a fire occurring on the other side of the partition.

In the majority of buildings the maximum fire resistance standard is one hour. This can be achieved fairly readily by most of the lightweight partition systems available, but particular attention must be paid to the perimeter details, where the partition connects to other building elements. Any gaps at this point will nullify the fire resistance. When considering the top edge of the partition, it must also be remembered that, if the ceiling is of the suspended type, a fire stop must be provided in the ceiling void along the line of the partition (see Ch. 20).

Where a higher standard of resistance is specified, the partition would have to be of a permanent nature, in which event much better gradings can be achieved. For instance, a 100 mm lightweight concrete block partition gives 4 hour protection.

Legislation to control the spread of fire dates back to 1666 and the Great Fire of London. In consequence, it is extensively covered in the London Building Acts and Part B of the Building Regulations, as well as in the Fire Precautions Act and Code of Practice 3 to which reference must be made in relevant circumstances.

16.3.4 Sound insulation

Lightweight partitions of any sort cannot be expected to provide any significant insulation against airborne sound. Any cracks or gaps in the partition (such as one must expect with movable partitions), would reduce such insulation as the partition possessed, practically to nil.

The magnitude of a sound is measured in decibels (dB), which is the ratio between the pressure of the perceived sound and the pressure of another sound. Thus, the unit can be used to express the level of sound in a room by relating it to a common standard, or it can be used to express the reduction of pressure between the sound source one side of, say, a partition and received sound pressure on the other side. The common standard reference point adopted is just below the threshold of hearing, and is set as zero on the dB scale. It is a logarithmic scale and, therefore, whereas a pressure of 10 times the standard is 1 bel or 10 decibels, a pressure of 100 times the standard is 20 decibels, and so on. The approximate values of background sound levels are:

Quiet conversation	30 dB
Conference room	50 dB
Normal conversation	60 dB
General office	80 dB

The insulation value of normal partitions is in the range of 25 to 45 dB. The lowest of these values relates to a half-glazed partition and the highest value to half-brick wall. Since the measurement is that of the ratio of sound pressure, it is incorrect to assume that a partition with an insulation value of, say, 30 dB will reduce the background noise level of a general office (80 dB) to that tolerable in the board room (50 dB). For this purpose a reduction more in the order of 45 dB is needed.

A partition insulates against sound by absorbing some of the sound energy in the effort required to make it vibrate. The more effort needed, the greater the absorption and the better the insulation. From this it will be seen that the mass of any sound insulating element is the characteristic from which its performance derives. This is why brick walls are better attenuators than single-glazed panels. If a very high standard is required, the mass must be further increased by the use of a thicker brick wall or one built of dense concrete blocks. In this case the ability of the structure to carry the load must be checked.

Where the partition is constructed *in situ* and finished by plastering, there should be no gaps to allow a noise leak, but prefabricated panels, with their many dry joints, need to be carefully fitted and resilient seals introduced (see Fig. 16.1).

In theory, discontinuous structures can achieve high standards of sound insulation. A discontinuous structure is one in which the surface on one side is not connected to the surface on the other side in any way. In practice this is almost impossible to achieve because of the physical constraints around the edges. If, for instance, the halves of a discontinuous partition are both fixed to the external wall, a path is immediately formed for a flanking transmission of sound which can virtually eliminate the value of the discontinuity in the partition.

Chapter 17

Specialised doors

17.1 Fire-resisting doors

Since the fire resistant quality of a door on its own has little
relevance to the way in which it would perform in a building, it must
be tested in conjunction with its frame and reference will be made in
this chapter to 'doorsets'. This is intended to indicate the door and
frame complete but not necessarily the standard package of door and
frame marketed by some manufacturers under this same name.

A fire-resisting doorset must contain a fully developed fire for a
period long enough to permit the escape of the occupants and the
fighting of the fire, therefore the resistant quality of a doorset is
expressed in terms of time. The actual duration of resistance will
usually be greater than the standard specified, since this latter does
not include the time taken for the fire to develop to the condition
where ignition of all exposed combustible materials has occurred,
(which is the definition of a fully developed fire).

Any doorset in the path of a fully developed fire must not:
1. become ineffective as a barrier due to collapse during a minimum
   period of time – this is referred to as its stability;
2. have or develop holes or gaps which allow the passage of hot
   gases or flames – this is referred to as its integrity;
3. become excessively hot on the exposed face, i.e. not more than
   180 °C above ambient at any point, or 140 °C above ambient
   overall – this is referred to as its insulation;
4. emit high levels of radiated heat where inflammable materials
   may be stored near the door.

The Building Regulations set minimum standards of integrity only but stability, insulation and radiation standards may be specified in certain applications.

In 1972 the British Standards Institution introduced a new test method, in BS 476: Part 8, which reproduced the real-life conditions by allowing pressure to develop within the test furnace. A direct consequence of this test method was to show that doorsets, previously considered suitable were, in fact, not satisfactorily resistant to the passage of flame, i.e. had a low integrity. The Building Regulations at the time tried to resolve this problem by referring to the test method in Part 8 as the recognised procedure but permitting the use of doorsets tested in accordance with BS 476: Part 1, provided that the test was carried out before 1973 on an identical doorset.

This dual standard has now been resolved. Appendix B to Approved Document B refers to BS 476 Part 22 and sets out in Table B1 the minimum standard of fire resistance for doors in specific positions. The old test method in Part 8 has now been replaced by Parts 22 to 24 and came into effect on May 29th 1987.

Part 20 gives general principles in the fire testing of building materials and structures and must be read in conjunction with Parts 21 to 24. In the main, the revisions relate to improvements in the techniques for measuring temperature, pressure and load distribution under test. This latter consideration, of course, does not apply to fire resisting doors.

Parts 22 and 23 are relevant to doors, the first deals with methods for determining the fire resistance of non-loadbearing elements of construction and the second deals with testing the contribution of intumescent seals to the fire resistance of timber doors.

The introduction of positive furnace pressure has made it necessary to eliminate any gaps in the test specimen through which hot gases could be forced, if the subject is to receive approval. With most structures this is solved by the use of fillers or sealants but with doorsets this creates a problem. There has to be a gap between the door and the frame to allow the door to open. The answer lies in the use of heat activated sealing systems, the most common of which is the intumescent strip. One, or two of these are fitted – preferably to the frame – and swell when the temperature reaches 200 to 250 °C, filling the gap around the door, thereby preventing the leakage of hot gas. The strips are preferred in the frame rather than the door edge to allow the latter to be planed in the normal way when hanging the door – or subsequently when easing it – without degrading the integrity of the doorset.

Fire resisting doorsets are rated according to their stability and integrity stated in minutes, thus: a doorset specified as 30/20 means that it resisted collapse for 30 minutes but allowed flames or hot gases to leak out at 20 minutes. The capacity of the door to provide insulation against a fire can also be tested but, as already noted, the Building Regulations only recommend minimum standards of integrity.

Many fire doors are hung on butt hinges and furnished with a latch or lock, and a door closer. All these require consideration. The hinges should have a narrow blade so that the intumescent strip is not covered, thereby restricting its action, and should have as light a mass as practicable, to reduce the amount of conducted heat. They must also, of course, resist the effect of fire and heat up to 850 °C. Locks and latches should be fitted as accurately as possible, to leave the maximum amount of door stile timber each side. It is also recommended by the Timber Research and Development Association (TRADA) that the lock case is coated with intumescent paint or paste. There are many types of door closer, fitted either to the face of the door or into the top rail or hanging stile and, to be certain that they will operate when required, they should be tested with the doorset.

Fire doors must not be fitted with any cabin hooks or a floor spring with a stand-open device. If the convenient use of the building requires the fire door to stand open, the Regulations permit the use of a fusible link or an electromagnetic device susceptible to smoke.

Preventing the escape of smoke round the door by sealing the gap not only enhances the fire-resisting quality of the door, it also eliminates any visual indication that there is a fire burning on the other side. For this reason, glazed vision panels are likely to be a regular feature of fire-resisting doors. In this respect there was a lack of definite advice because the relevant old Code of Practice, CP 153: Part 4: 1972 (now withdrawn), related to doorsets tested in accordance with BS 476: Part 1 and, as mentioned, these proved to be inadequate when tested in accordance with the newer Part 8. Figure 17.1 shows details of a typical fire resisting door and frame and includes a vision panel detailed to TRADA recommendations to satisfy that test.

Not only must the door and frame together receive consideration, but so also must the frame and wall junction. Many door frames are built in with a large gap between the frame and the brick jamb which subsequently may be covered by architrave mouldings. The accurate fit and fixing of these may achieve an adequate degree of integrity but to be certain that failure does not occur at this point the gap should be filled with plaster, mineral wool or intumescent mastic.

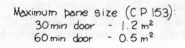

Maximum pane size (C P 153):
30 min door - 1.2 m²
60 min door - 0.5 m²

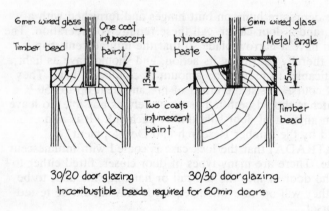

30/20 door glazing   30/30 door glazing
Incombustible beads required for 60 min doors

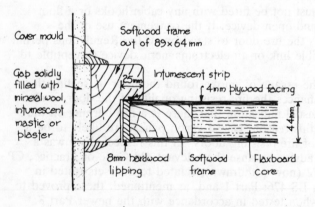

30/30 Fire-resistant doorset

Fig. 17.1 Typical fire door details

17.2 Sound-insulating doors

The doorsets used in normal building practice present very little resistance to the passage of sound between the spaces connected by the door. In most installations this is not important, but in certain situations it can be irritating, unacceptable or embarrassing.

There are two problems to solve in designing a sound-resistant door: the transmission through the door itself and leakage round the edges. The first is difficult to solve because, as already mentioned in connection with other constructions, the sound absorption of an element is proportional to its mass, but increasing the mass of a door to an amount which would provide a standard equal to the adjoining wall would make it very difficult to move the door to open it. The second problem, leakage round the edges, is readily dealt with by compressible seals, but careful installation of the frame or lining is also required to ensure that there are no gaps left around the perimeter of the doorset.

Sound-resistant doors are manufactured in either timber or metal

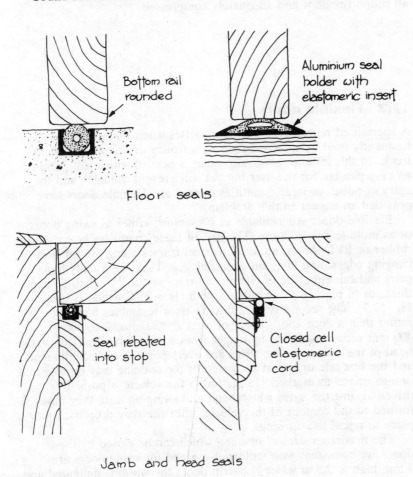

Fig. 17.2 Typical door seals

with flush facings and a variety of specially designed sound-insulating cores. The sound reduction for a timber door can be expected to be between 30 and 45 dB, depending on the thickness, and its weight between 12 and 30 kg/m². Steel doors can vary from 22 to 55 dB insulation value, depending on their construction, and their weight varies proportionally.

Perimeter seals are usually of the compressible tube type, as manufactured for draught exclusion (see Fig. 17.2). This aspect of sound insulation is particularly important; fitting such seals to an ordinary hollow-cored flush door will make a considerable difference to its sound-insulating properties, and therefore, care must be taken in its installation to make certain that the seals are both continuous all round the door and adequately compressed.

17.3 Flexible doors

A specialised requirement for doors arises when the opening is frequently used by someone pushing a trolley or driving a fork-lift truck. In this instance an arrangement is needed which will permit an easy passage for the user but yet will prevent draughts and the entry of noise, or large quantities of cold air. Flexible doors have provided an answer to this problem.

Flexible doors are available in two forms: either as swing doors or as multi-leaf strip doors. The first of these comprises heavy gauge rubber or PVC sheets, attached to steel frames along the top and hanging edges, and hung on spring hinges. They are usually hung in pairs and can either be of completely transparent PVC 7 to 10 mm thick, or of black PVC or rubber, with clear PVC windows (see Fig. 17.3). The second type of flexible door resembles a curtain rather than a door and is made of clear PVC strips, 200, 300 or 400 mm wide, suspended by hinges from a steel tube fixed across the head of the doorway. The strips are overlapped to exclude draughts and the first one or two at each side of the opening may be coloured orange or red as markers. In operation the vehicle is pushed or driven against the strips which bend and swing up until they have formed to the contour of the vehicle, after use they drop back into place to reseal the opening.

The maximum sizes of opening which can be closed by these doors are consistent with industrial use and for swing doors are 3.5 m high × 3.5 m wide. For strip doors the width is unlimited and the maximum height is 6.0 m.

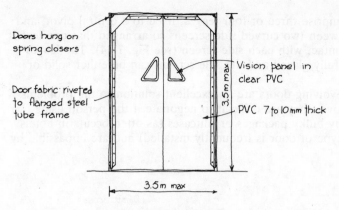

Doors hung on spring closers

Door fabric riveted to flanged steel tube frame

Vision panel in clear PVC

PVC 7 to 10mm thick

3.5m max

3.5m max

Flexible swing door

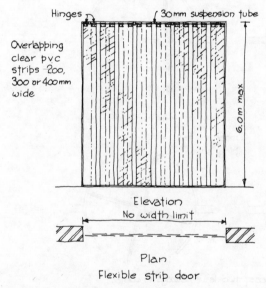

Hinges

30mm suspension tube

Overlapping clear pvc strips 200, 300 or 400mm wide

6.0m max

Elevation
No width limit

Plan
Flexible strip door

Fig. 17.3 Typical flexible doors

17.4 Revolving doors

Whenever a normal external door is opened, more than just the person entering or leaving the building is admitted; large quantities of cold air (or hot air in some parts of the world) can find their way in. Revolving doors are a means of reducing this volume of air while still allowing people to move freely in and out.

They comprise three or four leaves, fixed to a central pivot, and rotating between two curved side screens so arranged that one leaf is always in contact with each side screen (see Fig. 17.4). The leaves are usually fully glazed and the side screens can be either solid or glazed.

While revolving doors are an excellent solution to the problems outlined above, they are difficult to negotiate if the person using them has any bulky packages or suitcases (as often occurs in hotels where this type of door is frequently installed) and are impassible by

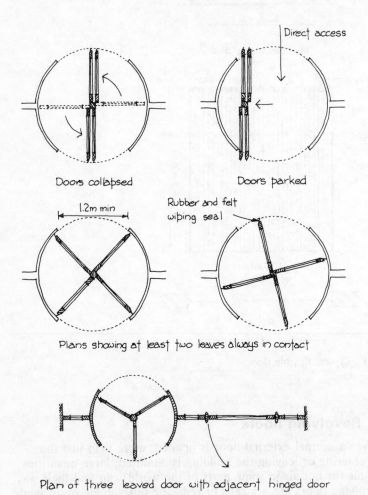

Doors collapsed

Direct access

Doors parked

1.2m min

Rubber and felt wiping seal

Plans showing at least two leaves always in contact

Plan of three leaved door with adjacent hinged door

Fig. 17.4 Revolving doors

a person in a wheelchair. They also are not suitable as fire exits. For these reasons it is common practice to provide a normal hinged door, adjacent to the revolving doors, for occasional use.

Alternatively, the leaves of the revolving door can be hinged at the centre, and the pivot fixed to a trolley, so that they can be collapsed and slid to one side to provide a straight-through access. This is still not suitable as a fire exit, but can solve the problems of access with parcels or wheelchairs.

The door units are made of either timber, usually hardwood, or aluminium, usually anodised, and have overall diameter of 1.8 to 2.1 m although a three-leaved door can be as small as 1.5 m diameter.

Chapter 18

Wall finishes

18.1 Choosing a wall finish

In this chapter, the word 'finish' is taken to mean the final permanent surface of the wall construction, whether or not it is the decorative surface as left at contract completion. Thus the term does not embrace any of the applied paint or wallpaper techniques, but can include decorative finishes where they are an integral part of the finishing process.

There are very many ways in which a wall may be finished, but they are all readily divisible into 'wet' or 'dry' and this fundamental distinction greatly affects the choice to be made.

Wet finishes are always made up and applied on site. Some dry finishing materials arrive on site in stock sizes and have to be cut and fitted, other dry finishes are either completely made up away from the site and fixing is the only site operation, or they are applied to the units which make up the wall when they are being manufactured and hence no site work is involved at all. The methods of on- or off-site finishing possess advantages and disadvantages similar to those applying to other materials and constructions in the building fabric and very much affect the final selection.

One of the most significant differences between wet and dry finishes is the time factor. All wet finishes must be thoroughly dry before the surface can be considered stable or inert and this can mean quite a long delay in the building programme. Since the finishing trades are always employed towards the end of the contract period, there is little opportunity to absorb any deviation from the building programme and with drying times largely subject to the vagaries of the weather, much anxiety can arise as the completion date looms nearer.

As well as the time factor, the other differing characteristics which influence the choice of finish are: accuracy of the structure, the accommodation of service runs, the extent to which the appearance is to be alterable and the architectural character desired.

Dry finishes of the prefabricated type require a high degree of accuracy of the background to which they are applied, whereas wet finishes can be used to correct inaccuracies to limited extent. Fixing dry finishes such as timber battens and plasterboard can be a much slower process if irregularities in the wall face make a great deal of trimming or packing of the battens necessary.

Finishes applied away from the site create difficulties over the concealment of service runs such as electrical cables to switches and sockets. In most cases these are routed through cable ducts built into the walling or partitioning system (see Ch. 16). This, however, can result in electrical points being placed where the system permits, rather than where the user requires. Most finishes applied to the walls *in situ* allow cable runs to be installed first and subsequently concealed by the finish.

Some finishes, such as plaster, are not intended to be left as applied and receive a decorative coating. This can and indeed must be renewed regularly to maintain a good standard of appearance. For many users this necessity is an advantage, since it affords the opportunity to change the decorative scheme. Other finishes, such as wall tiling, can be changed if required, but the durability of the material and the cost of the operation tend to discourage any regular alteration. Thus these finishes must be selected for positions where a static appearance is acceptable. Similarly, integral finishes cannot be changed, they can only be overlaid by another finish if a different appearance is essential and thus must also be confined to rooms where permanence of decor is preferred.

18.2 Types of wet finish

A wet finish is invariably one of the various types of *in situ* plaster systems available. In this connection the term plasterwork is intended to cover the coating of internal walls and ceilings. Similar work carried out externally is usually called rendering but, confusingly, the first coat of a plaster treatment is also called the render coat.

The plaster can be made up on site either from a mixture of an aggregate, usually sand, and a binder such as cement, lime or gypsum; or from a pre-mixed proprietary dry material which only requires the addition of water. The majority of plasterwork uses the on-site mixture; pre-mixed plasters are usually more expensive but are used where the finish must possess special characteristics such as sound absorption or X-ray protection.

Wall plastering is almost always applied in two coats, the render coat and the setting coat. In good quality work a three-coat system of render, float and set may be used, but where the background is very true a 'thin-wall' single-coat plaster can be applied. An exception to this is projection plaster, which is a gypsum-based material sprayed onto the wall to a thickness of 13 mm in one coat and trowelled to a true face.

The first undercoat, or render coat, must be appropriate for the background to which it is applied, both in its thickness (see Fig. 18.1) and in its strength (no plaster coat should be stronger than the surface to which it is applied). Cement-based undercoats (one part Portland cement to three parts sand) require a rigid well-keyed background such as brick or concrete and usually are slightly cheaper than gypsum-based undercoats. The latter material (one part browning grade plaster to two parts of sand – depending on the background) produces a superior finish, adheres better and does not shrink on drying – as does a cement-based undercoat.

The setting coat must set at a rate which allows the plasterer to work it to a true and level surface, and must then provide the desired hardness of finish. Gypsum plaster is generally used for this purpose because it goes firm in $\frac{1}{2}$ to 1 hour and can then be dampened and trowelled up to a close dense surface. It also produces an inert craze-free face. A gauged cement setting coat (one part cement to three parts sand gauged with between one quarter to one part lime putty), can be applied to a cement-based render coat where an exceptionally heavy duty finish is needed but the finished surface is alkaline, which limits the range of decorations which can be applied, and it is usually crazed because of drying shrinkage.

Reference is made in Fig. 18.1 to classes of gypsum plaster; this is in accordance with BS 1191: 1973 *Gypsum Building Plasters* which divides gypsum plaster into the four following classes:

Class A: Hemi-hydrate calcium sulphate plaster. This is the mineral gypsum with three-quarters of its water of crystallisation driven off. Commonly called plaster of Paris, it sets very quickly and is limited in use.

Class B: Retarded Hemi-hydrate Plaster. This is plaster of Paris with additives to retard the setting time. There are two sub-grades: *Type a*: 'Browning' and 'metal lathing' grades which, mixed with sand, are used for undercoats. *Type b*: 'Wall finish' used for final coats and 'board finish' used for one coat work on plasterboard.

Background	Gypsum based plaster						Cement based plaster				Premixed plaster			
	Render Class B	Thick (mm)	Float Class B	Thick (mm)	Set Class B	Thick (mm)	Render	Thick (mm)	Set	Thick (mm)	Render	Thick (mm)	Set	Thick (mm)
Dense brickwork	1Br:2S	11	-	-	F neat		1Pc:3S	11	Keenes	3	Br	11	F	
Average brickwork	1Br:2·3S	11	-	-			1Pc:¼L:3S	11	Class B or C	2	Br	11	F	
Light. conc. block	1Br:2S	11	-	-		2	1Pc:¼L:3S	11	Not recommended		Br	11	F	1·5
Dense concrete	1Bo:1½S	8	-	-			Not recommended				Bo	8		
No-fines concrete	1Br:2S	11	1Br:2S	6			1Pc:1L:5·6S or 1Mc:4½S	11	Class B or C	2	Br	11		
Aerated concrete	Not recommended						Not recommended				Not recommended			
Plaster board	-	-	-	-	Bf neat / F neat	5 / 1·5	Not recommended				Bo	8	F neat	2
Exp. metal lath	1ML:1½S	8	O* 1ML:2S	11	F neat	2	Two coats*	11	B or C	2	ML	8	F neat	1·5

NOTES:

* The first coat on expanded metal lath is trowelled through the mesh

O* First coat 1Pc:¼L:3S
 Second coat 1Pc:½L:4S
 or
 1Mc:3S

KEY:

Bf : Board finish plaster. Class B.
Bo : Bonding grade. Class B. a3
Br : Browning grade. Class B. a1
F : Finishing grade. Class B.
HBr : Haired browning grade. Class B. a1
L : Non-hydraulic lime putty
ML : Metal lathing grade Class B. a2
Mc : Masonry cement. BS 5224:1976
Pc : Portland cement. BS 12:1978
S : Sand. BS 1199:1976

Fig. 18.1 Wet finishes and backgrounds

Class C: Anhydrous gypsum plaster. This is the mineral gypsum with all the water of crystallisation removed. Initially its sets very quickly but it can be retempered by the addition of more water and the second set is slow enough to provide a finishing plaster suitable for gypsum or cement-based undercoats.

Class D: Keene's cement. This is also anhydrous calcium sulphate (like Class C) and is used as a finishing plaster to produce a very hard surface such as is required on the back wall of a stage to reflect sound or on the walls of a squash court to withstand the impact of the ball.

18.3 Types of dry finish

Dry finishes comprise some form of sheet material fixed to the wall, or to battens on the wall, either with or without an integral decorative finish. The joints between the sheets may be left showing, covered with a trim or stopped up. The sheet material most commonly employed for this purpose is plasterboard or wallboard – this particular application is generally referred to as dry lining – but plain hardboard and plywood, faced or veneered hardboard and plywood, softwood or hardwood boarding and timber panelling are all used as dry finishes.

Dry lining wallboard has a tapered edge to allow the joint to be filled flush with the face (see Fig. 18.2). They are either fixed to battens which are plugged and screwed, cut nailed or shot fired to the wall surface, or they are secured to the wall by plaster dabs. With the latter method, the first operation is to attach 50 × 75 mm 'dots' of bitumen impregnated fibreboard to the wall with board-finish plaster in three horizontal rows, one just above the bottom edge of the boards, one just below the top and the third at mid height. Spacing of the dots is 450 mm for 900 mm wide boards and 400 mm for 1200 mm wide board. They are carefully aligned with a straightedge, both horizontally and vertically. When the dots are firm, board finish plaster dabs are applied to the wall in rows to suit the board size and the boards are pressed onto the dabs until they stop on the dots to give a true surface. Once in place they are temporarily secured with double headed nails which are removed once the plaster dabs have set. The joint is then partly filled with a specially made joint filling plaster, a 44 mm wide paper tape is pressed into it and the joint flushed up with a second application of filler. When the filler has set, a thin layer of joint finish plaster is applied in a band 200 to 250 mm wide with the edges feathered off with a sponge.

A dry lining of plain hardboard is the cheapest method used but is not a very satisfactory finish. The board should be wetted on the

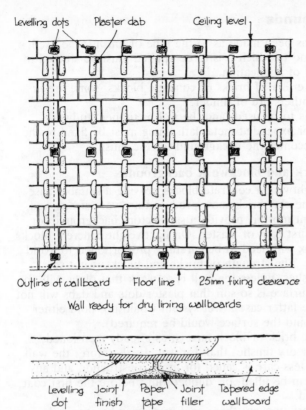

Levelling dots   Plaster dab   Ceiling level

Outline of wallboard   Floor line   25mm fixing clearance

Wall ready for dry lining wallboards

Levelling   Joint   Paper   Joint   Tapered edge
dot   finish   tape   filler   wallboard

Plan of joint

Fig. 18.2 Dry lining with gypsum wallboard

back before fixing so that it is in a slightly expanded condition when
fixed and tightens when it dries. This is to restrict movement of the
board with humidity changes, but even with this precaution a
hardboard lining can be readily recognised by the waviness of the
surface. Plain plywood does not suffer from this problem, but is
more expensive.

Decorated sheets can be faced with a wide variety of materials
but mainly they are hardwood veneers, plastic laminates or fabrics
such as hessian or vinyl cloth and are fixed with adhesive, screws
and caps or with clips. Joints are either left showing or covered by a
metal trim.

Timber boarding is usually fixed to battens, but panelling may be
plugged and screwed directly to the wall. In either case, the backs of
the materials must be ventilated and care taken to avoid any
possibility of dampness, otherwise there is a risk of dry rot
developing.

18.4 Backgrounds

Once a decision has been made as to the type of finish to be used, then the background to which it is to be applied must be studied to decide the method of application.

Most walls are either of bricks or concrete blocks, some are *in situ* concrete and others are of timber.

If a wet finish is to be used, and the wall is timber studwork, then a wallboard or expanded metal lath lining must be fixed to the studding, which becomes the background for the finish.

18.4.1 Brickwork and blockwork backgrounds

Clay bricks and lightweight concrete blocks provide a suitable base for plastering. Some clay bricks now produced are very strong and, when laid in cement mortar, provide a satisfactory background for an equally strong first coat of plaster. On lightweight concrete blocks the render coat mix will need to be carefully proportioned to avoid it being too strong. The strength of bricks and blocks is not so critical for dry finishes, unless the wall is so weak that plugs will not hold in it, or the surface is so soft that plaster dots and dabs will not adhere to it (in the latter case a wet system would not hold either and treatment to bind the surface would be required).

Accuracy of the brick or block face is, as already mentioned, more important for dry finishes than for wet. If, therefore, the wall is to be plastered, less care is called for in its construction and any local depressions can be filled, or 'dubbed out', before the first coat of plaster is applied.

If the wall is very uneven, three coat work may be required – which offsets any cost savings resulting from the lower standard of accuracy in the wall. For a dry finish to be satisfactory, any irregularities in the wall must be eliminated by the preparations made to receive the lining material. If the preparation is to batten the wall then these must be fixed so that their faces are true, by means of packing out over the low spots or cutting away (scribing) the batten over the high spots, with 40 or 50 battens to deal with (as can be the case in an average room) the preparation can take a long time. The dot and dab method of dry lining can accommodate wall inaccuracies much more easily, since all that is required is to ensure that the dots are aligned so that no high spots in the wall protrude in front of the plane of the back of the wallboards.

Key and suction are two factors of the background which affect wet finishes only. Most brick and block walls provide an adequate key, especially if the joints are raked back rather than carefully flushed up. Lightweight concrete blocks are scored on their surface to provide the mechanical key required to hold the weight of the plaster. Suction by an absorptive background can cause the water to plaster ratio to be changed before the plaster has set; this either can increase the density of the mix, thus producing a stronger plaster than required, or it can remove the water so rapidly that there is insufficient time to work the material properly into the keys, and adhesion is permanently weakened. Suction can be reduced by wetting the surface but this can lead to excessive drying shrinkage. It is better to apply a polyvinyl acetate (PVA) emulsion bonding agent, or to include a water-retaining additive in the mix. Some ready-mixed plasters contain a cellulostic material for this purpose but, on site, fat lime putty can be used.

A problem which can be encountered in brickwork, and occasionally in blockwork, is the formation of efflorescence – salts which are contained in the bricks or blocks which migrate to the surface of the wall as it dries, there to form a white crumbly deposit. An advantage claimed for dry finishes generally is that they are unaffected by this. As far as wet finishes are concerned, provided that the wall is thoroughly dry, so that no more efflorescence will occur, and such salt deposits as have formed are completely removed with a stiff brush (no water), there should be no further trouble. If the background is not thoroughly dry then the render coat should be selected to provide a barrier to the salts; in this respect cement-based coats are better than those using gypsum plaster.

18.4.2 In-situ concrete backgrounds

Where a dry finish is specified on a concrete background the only matter for concern is the method of fixing. Modern shot-fired fixings and percussion drills for plugs can easily handle most problems of mechanical fixings into concrete, or purpose-made fixing points can be provided in the concrete when cast. Success of the dot and dab method of dry lining is more reliant on the nature of the concrete surface and will be subject to the same considerations as wet finishes.

There are basically three types of concrete: dense, lightweight and no-fines, and each requires different treatment where a wet finish is specified.

Dense concrete cast against smooth formwork offers very little support for a plaster finish; there is virtually no mechanical key, the surface has a high suction and movement due to drying shrinkage can occur. Nothing can be done about the last of these, shrinkage, except to delay plastering until the concrete has thoroughly dried out. A mechanical key can be obtained by hacking the surface or by brushing it after it has been kept soft by the use of concrete retarders painted onto the formwork. Alternatively the face can be formed into dovetail grooves by the use of special linings fixed inside the formwork. This last method is the most reliable.

An alternative pretreatment to dense concrete is to coat it with a

PVA bonding agent. This has the added advantage that it also controls the rate of suction.

All pretreatments are intended to provide a surface which is suitable for normal plaster systems. However, a satisfactory finish can be achieved with little or no pretreatment by the use of a pre-mixed lightweight bonding plaster undercoat followed by a lightweight finishing plaster or, where the concrete is sufficiently true, a single skim coat of neat board-finish plaster (Class B).

Lightweight concretes, either aerated or those containing lightweight aggregates, must be thoroughly dry before plastering to ensure no shrinkage of the background. The efficiency of the mechanical key and the degree of suction to be expected varies widely according to the product and most manufacturers produce a recommended plastering specification which should be followed.

No-fines concrete provides an excellent mechanical key, it has a low to moderate drying shrinkage and, with a careful selection of the aggregate, can be given the right degree of suction to make it a suitable background for plastering.

18.4.3 Plasterboard backgrounds

All plasterboards consist of a core of gypsum plaster faced each side with heavy paper and are described in BS 1230: Part 1: 1985. They are manufactured in three basic forms: wallboard, baseboard and lath. Wallboard has one face lined with an ivory paper for direct decoration and the other with grey paper which is intended to receive a plaster finish. The most commonly stocked size is 9 mm thick × 1200 mm wide × 2400 mm long, but 12.5 mm thickness, 600 and 900 mm widths, and 1800 and 3000 mm lengths are also available. Baseboard is faced on both sides with grey paper for plastering and is only made in one size: 9 mm thick × 914 mm wide × 1200 mm long. Lath is similar to baseboard but in smaller sizes: 9 or 12.5 mm thick × 405 mm wide × 1200 mm long. In addition to these the following are also produced: taper-edged wallboard for dry lining; gypsum plank, which is 19 mm thick and used where sound insulation or high fire resistance is required; insulating plasterboard, which has a bright metal veneer of low emissivity on one or both faces for thermal insulation; perforated or slotted plasterboard, used in conjunction with a glass fibre or mineral wool backing, for sound absorption; water resistant plasterboard, which has a specially treated core and water-resisting paper; insulating sarking board, which is water-resistant plasterboard lined with aluminium foil and used on roofs; and plastic-faced plasterboard used for predecorated dry lining.

Fixing should be by galvanised plasterboard nails, 32 mm long for 9 mm boards and 38 mm long for 12.5 mm boards, driven just below the surface, 13 mm in from all edges at 100 mm centres and at 200 mm centres across the centre of the board. The boards should be fixed with their long dimension running across the supports because the board strength in this direction is about twice that across the board, and it should be arranged that the ends of all boards butt together on a support, preferably staggered.

The finished face is true enough to require only a single skim coat of neat board finish plaster but a better standard can be obtained with two-coat work using a pre-mixed lightweight plaster.

18.4.4 Metal lath backgrounds

A very wide range of expanded metal sheeting is produced. The one generally used as a plastering background is plain expanded steel, which should be either galvanised or painted with bitumen. Metal lath, with integral or attached ribs to give greater strength, is also available.

Metal lath is fixed with large headed galvanised nails, and edges of adjacent sheets should be overlapped at least 25 mm at supports. Where the supports are wide the lath should be spaced away a short distance with rods or strips laid along the support to provide room for a satisfactory plaster key.

Three coats of plaster are essential for a satisfactory finish and gypsum plaster is preferable to cement. The render coat and the float coat should be in Class B metal lathing grade plaster with a Class B finishing grade plaster as the setting coat.

18.5 Wall, floor and ceiling junctions

Besides any aesthetic considerations of skirting and cornice design there are certain practical reasons for the selection of these and other trims. Any wall face is subjected to physical abuse along its bottom edge, due mainly to chair legs and floor cleaning machines and methods. Skirting boards have always been provided to withstand this. Cornices and cover moulds at junctions are a means of concealing cracks which inevitably develop where dissimilar materials meet.

Figure 16.1 shows the use of metal cover strips, plaster cornice and timber skirtings in conjunction with prefabricated partitioning. Their use in wet and dry finishes on permanent walls is very similar (see Fig. 18.3).

Cracking of plaster finishes on plasterboard can be controlled by reinforcing the background with jute scrim 90 mm wide bedded in neat board plaster over all joints and at wall-to-ceiling and wall-to-wall junctions. Similar control can be achieved in plaster finishes on solid backgrounds by fixing expanded metal lath over the junctions of any changes of material, such as brick to concrete lintel, or brick to timber wall plate.

One type of cracking is unavoidable and occurs along a horizontal line 400 to 700 mm below the ceiling in upper floors of pitched roof buildings. This is caused by the strapping of the roof timbers carried down inside the wall, as required by the Building Regulations, moving the top courses of blockwork as the timbers shrink or twist, thus disturbing the plaster finish. The only treatment is to cut out and stop up the crack once the roof has settled down.

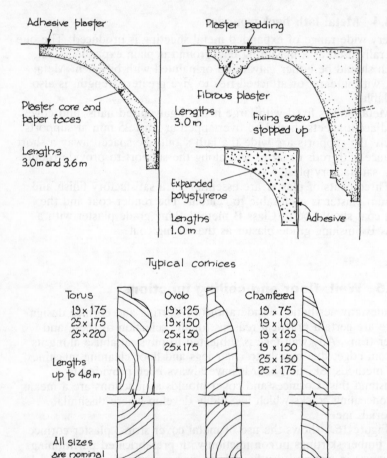

Fig. 18.3 Typical skirtings and cornices

Chapter 19

Floor finishes

19.1 Types of finish

With one exception, the finishes described in this chapter are applied
to a floor which is structurally complete, and can be grouped into
four classes:

Timber flooring:
Included in this class are all the decorative
timber floor finishes which, when laid, are
intended to be seen. Thus, for the purposes of
this chapter, timber boards and sheets which
act as the floor structure are excluded. The
exception referred to above is in this class and
is hardwood strip flooring which can be laid
direct to the floor joints and thus provides both
floor deck and finish.

Sheet flooring:
This covers all the materials which arrive on
site in a roll of predetermined width and
variable length. They may cover the room in
one piece or there may be a few joins.

Tile flooring:
In this class the material is delivered to site in
pieces which have all the dimensions fixed.
These can vary from as small as 50 × 50 mm
up to tiles 1 m square.

Jointless flooring:
These floorings are applied to the floor in a
plastic form either ready mixed or prepared on
site, and are worked over to produce the
specified finish.

The characteristics and advantages of the materials within each class are very wide and varied.

Generally, timber flooring produces a finish with a warmth of colour and natural variation which in the right setting can be very attractive and it can also be a very hard wearing surface; sheet flooring provides the opportunity to introduce pattern and texture in the floor and, with the small number of joints, it can also be a hygienic finish; tile flooring is much more economical in material than sheet flooring because less wastage is involved in fitting to the room size and cutting round fixed objects; jointless flooring is the most hygienic finish of all and can even be taken up the walls as a skirting to provide a surface which can be kept clinically germ free.

19.2 Timber flooring

A major problem with timber flooring is the movement to which it is subject following changes in humidity. The timber is always dried to a specified moisture content by the timber merchant, but if this fails to match the ambient humidity of the building, dimensional changes will occur, leading either to gaping joints, or to buckling or lifting of the flooring. Unfortunately, the humidity can vary from time to time causing the timber to swell or shrink accordingly. This is particularly true of new buildings (in which the majority of the flooring is laid) and much care must be exercised to see that the building is dried to a tolerable level by heating and ventilating (heating alone is ineffective).

Where the heating system is closely associated with the flooring – as in underfloor heating coils or ceiling panels – it is recommended that this is run for at least three weeks before the timber is laid. The timber must be dried to below 10 per cent moisture content, preferably 8 per cent, and a suitable species must be selected which can withstand these conditions. Chapter 20 in *Advanced Building Construction*, Vol. 1, deals with timber for building and Table 20.2 therein lists the common hardwoods, their characteristics and their uses.

From the point of view of colour. Table 19.1 below sets out appropriate timbers which produce specific flooring colours but it must be recognised that all these will gradually change; in most cases it is a toning down or deepening of colour, but bleaching by sunlight can also occur.

Whichever timber is selected, the form in which it is used can be strip flooring, wood block flooring, parquet flooring or wood mosaic flooring.

In all types of hardwood floor, adequate sealing and polishing is essential to develop the beauty of the timber and subsequent regular maintenance (particularly in the first year) is needed to ensure a long life.

White	Light yellow	Browns	Reds
Selected Maple Hornbeam	Maple Oak Loliondo Beech Ayan Canadian Birch Idigbo Afara	Teak Iroko Afrormosia	Afzelia Jarrah Danta Missanda Muhimbi Rhodesian Teak Keruing Sapele

Fig. 19.1 Table of flooring timbers

19.2.1 Hardwood strip flooring

Hardwood strip flooring is composed of narrow tongued and grooved hardwood boards, secret nailed to timber joists at centres not exceeding 450 mm, or to 50 × 50 mm impregnated timber battens, fixed to a concrete floor with metal clips. The nominal widths of board are 50, 63, 75, 88 and 100 mm, the finished face width is from 13 to 19 mm less than the nominal width depending on the profiles (see Fig. 19.2). Nominal thicknesses are 22, 25, 32 and 38 mm with 5 or 6 mm off for the finished size.

Strip flooring is generally laid parallel to the length of the room for greater economy and better appearance, which means that, in the case of a timber floor, the joists must be arranged across the narrow dimension of the room.

Expansion of the flooring must be anticipated and due allowance made, either by leaving the flooring well clear of the walls below a thick skirting, or by incorporating a compressible cork strip all around the perimeter. It is also good practice to leave an open joint every fifth or sixth board, to allow additional tolerance for the initial movements that can take place.

19.2.2 Wood block flooring

The most common nominal sizes of wood blocks are 225 × 75 mm or 300 × 75 mm and 25 mm thick. The thickness can vary down to 19 mm or up to 38 mm. Most blocks are milled with a tongue on one long edge and one end and a groove on the other edge and end. This tonguing and grooving is kept towards the bottom of the block because the wearing thickness of the block is that portion above the tongues and grooves.

The patterns to which blocks are laid are herringbone, basket weave, double herringbone and brick bond (see Fig. 19.2). Herringbone pattern flooring can be laid either square with the walls, or diagonally to them, and is the most popular, because the blocks are evenly disposed at right angles to each other and any

172

movement of the timber (which is greatest at right angles to the grain) is uniformly distributed. As with strip flooring, allowance must be made around the perimeter of the floor for this movement, either by a generous gap or by an expansion strip. If a cork strip is inserted it is usual practice to lay two or three rows of blocks around the edge of the floor parallel to the walls against which the cork is fitted.

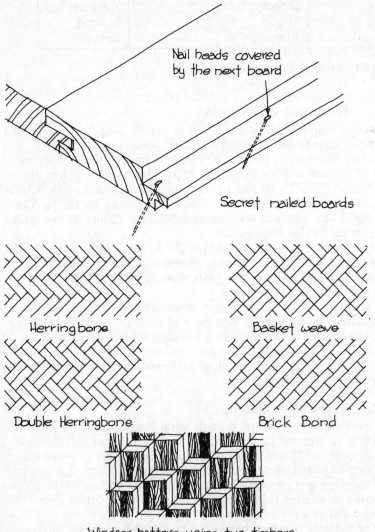

Nail heads covered by the next board

Secret nailed boards

Herringbone

Basket weave

Double Herringbone

Brick Bond

Windsor pattern using two timbers

Laying patterns of wood blocks and parquet

Fig. 19.2 Wood floor finishes

Wood blocks can only be laid on a concrete floor. This must be accurately levelled by a screed, thoroughly dry and have a damp-proof membrane within it, if it is a ground floor. The blocks are bedded onto the floor in an adhesive, which may be hot or cold. In many cases this adhesive is bitumen based and could be thought to provide a damp-proof barrier but it is unreliable in this respect and separate, preventative measures against rising damp should be provided.

19.2.3 Parquet flooring

Parquet flooring, when laid, looks rather similar to wood blocks, the main difference being that the blocks are longer, narrower and quite a lot thinner; a typical size is 250 × 50 × 6 mm. The parquets are not tongued and grooved but may have a dovetailed recess worked along the lower edges into which the adhesive is squeezed when laying, to improve the bond to the floor.

Like wood blocks, parquets are stuck down to concrete floors but they can also be laid over timber floors and in this case the floor is first covered with plywood (unless flooring grade chipboard or blockboard has been laid over the joists) and the parquet blocks are pinned in each corner as well as being stuck.

Laying patterns are usually herringbone or basket weave, sometimes using timbers of differing colours to produce a design. Another form which achieves a very attractive floor is the Windsor pattern which also uses two differing timbers as well as parquets with diagonally cut ends (see Fig. 19.2).

19.2.4 Wood mosaic flooring

Again looking similar to wood block flooring, but very much smaller, wood mosaic flooring consists of 'fingers' of hardwood 100 to 115 mm long × 25 to 29 mm wide and 6 to 10 mm thick, arranged in herringbone or basket weave pattern and mounted on aluminium foil, bituminous felt or a netting, to form panels about 450 mm square. The panels are stuck to a concrete or timber sub-floor with an adhesive recommended by the manufacturer of the panel.

19.3 Sheet flooring

There is an enormous range of types, colours, patterns and qualities of sheet flooring available, but mainly they can be divided into: linoleum, rubber, vinyl (polyvinyl chloride) and carpet.

19.3.1 Linoleum flooring

Linoleum is the oldest of the flooring materials in use today and was developed out of varieties of 'oil-cloth' and a material called 'kamptulicon'. This latter flooring was made from cork dust and softened indiarubber cemented onto a canvas backing. In the 1850s Frederick Walton tried substituting oxidised linseed oil for the very expensive indiarubber and produced linoleum, which name he based on the Latin words for its chief ingredient: linum (flax) and oleum (oil).

Linoleum is now made from cork powder, wood flour and pigments bound with oxidised linseed oil, mounted onto a jute canvas backing. It is produced in many types and grades, the most common being: plain linoleum – uniformly coloured material where the colour passes right through to the backing; printed linoleum – patterns applied to the surface; decorated linoleum – where the linoleum mixture is carefully adjusted to give marbled, jaspé, granite or moiré effects; inlaid linoleum – a patterned flooring of separate areas of linoleum mixture made up into one sheet; cork carpet – in which cork granules are substituted for the cork powder.

The standard size of roll is 1.83 m wide × 23 to 28 m long and 2.0, 2.5, 3.2, 4.5 and 6.0 mm thick. The 4.5 and 6.0 mm thicknesses are usually supplied in shorter rolls of 11.5 to 14 m.

Linoleum can be laid on almost any type of sub-floor, but each requires special treatment to achieve the best results. It is bonded to the sub-floor with approved adhesive and when so bonded it is not necessary to allow the unrolled sheet to lay loose for a period to 'relax', as used to be the practice.

19.3.2 Rubber flooring

Both natural rubber and synthetic rubber sheet flooring is available, but the latter is the more commonly used, since it possesses superior qualities.

Natural rubber flooring is made by vulcanising the latex or sap from the rubber tree, by mixing sulphur and lead carbonate with it and heating the mixture. Synthetic rubber flooring is made from a range of chemicals which include chlorosulphonated polyethylene, styrene-butadiene copolymers and butyl and nitrile rubbers (see Ch. 21).

Both types of rubber flooring will last a long time and often fail because of wrong maintenance rather than normal use. Synthetic rubber offers a greater resistance to abrasion than natural rubber and does not craze and go brittle if regularly exposed to sunlight, as does its natural equivalent. Both types are susceptible to cleaners and polishes which contain petrol, paraffin, turpentine or similar spirits which can soften the surface of the flooring. They also are affected by oil and grease.

Natural and synthetic rubber flooring is manufactured in the same range of sizes of 3.5 and 6 mm thick, 900 mm wide in rolls up to 30 m in length. Sponge backed rubber is also manufactured to give even greater resilience to the floor, and consists of a solid rubber sheet 1.6 mm thick mounted on a sponge rubber backing 5 mm thick, or 3 mm rubber on a 3 mm sponge backing. This sponge rubber flooring can be used as a finish on concrete suspended floors as a means of providing resistance to the transmission of airborne and impact sound."

19.3.3 Sheet vinyl flooring

Work on producing flexible vinyl flooring began during World War II when the British Admiralty commissioned researchers to produce a substitute for the linoleum which had been used in battleships, but which was becoming scarce as well as proving to be not fully fire resistant. It is made from polyvinyl chloride, copolymers of vinyl chloride, or blends of these as binders, with white china clay and calcium carbonate as the fillers (see Ch. 21).

This flooring is extremely durable; its very close texture prevents any penetration by grit or dust particles, which could abrade the surface. It is completely impervious and offers high resistance to acids, alkalis, grease, solvents, fire, electricity and indentation. It is affected by a live cigarette butt ground into it. There are very few floor coverings that would not show some adverse reaction to this abuse. In the case of vinyl flooring the resultant discoloration can be removed with fine steel wool or a nylon scouring pad, but surface damage would remain.

The base mix of the material is white, therefore pigmentation of the flooring is extremely easy and the resulting colours are clear and vibrant. It is also possible to produce a translucent vinyl flooring into which is set a submerged pattern of vinyl chips or metallic particles.

Vinyl sheet is produced in four thicknesses: 1.6, 2.0, 2.5 and 3.2 mm in three widths: 900, 1350, and 1800 mm and in rolls 11 to 28 m long.

Like most floor coverings it should be stuck down to the sub-floor with an adhesive recommended by the manufacturer (some adhesives can cause staining or can extract the plasticiser, causing shrinkage). Being flexible, vinyl sheet flooring will readily assume the profile of the sub-floor and this, therefore, must be carefully levelled. Screeded floors should be checked for trueness and any irregularities corrected with a latex cement floor levelling compound, also their moisture content must not exceed 5 per cent. Tongued and grooved softwood timber boarded floors are usually covered with plywood or hardboard to eliminate any curl in the boards but chipboard or blockboard floors should be satisfactorily flat as laid. All timber floors, especially ground floors, must be well ventilated, because the vinyl sheet, when stuck down, is virtually vapour proof and thus can trap moisture in the sub-floor, thereby actively encouraging dry rot.

Where a jointless surface is required, as in a hospital, and the area is greater than can be covered by the width of the roll, adjacent sheets can be welded together either by being butted and heated or, more often, by running a welding strip of the same material into a small gap left between the sheets. Another flooring requirement of hospitals, and certain other users, is that it must be anti-static. In certain conditions a static electrical charge can build up on vinyl floors with undesirable consequences. To meet this, an anti-static vinyl flooring is produced containing various additives, such as carbon, to make it electrically conductive.

19.3.4 Carpets

A vast range of types, materials, colours and patterns of carpet are now manufactured within an equally large range of prices. Selection must take into account performance, appearance and cost, and it must be realised that these are not always directly related – a carpet can retain its appearance but could loose its resilience or it can resist wear but deteriorate in its appearance. A cheap carpet may offer a greater resistance in a particular situation than a more expensive one.

The materials used in manufacture are wool, a number of manufactured fibres, vegetable fibres and mineral fibres either on their own, but more often in a mixture proportioned to take advantage of the combined characteristics of each type of fibre.

Wool is very resilient, fast to dye, it has a good resistance to wear and is self extinguishing when burnt. It is not often used on its own but is usually combined with nylon or one of the other synthetic fibres in varying proportions, the best of which being 80 per cent wool and 20 per cent nylon (referred to as an 80/20 carpet).

Nylon has a very high strength and abrasion resistance, but is not so resilient as wool so is better as a short pile carpet. It used to be prone to static charging and showing dirt, but these problems have now been largely overcome and many more carpets are now being woven with a 100 per cent nylon pile.

Acrylic is nearer to wool than any of the other synthetic fibres but has a higher resistance to abrasion than wool. In most other respects it is very similar to nylon.

Viscose rayon is the cheapest of the synthetic fibres and the most inferior.

Polypropylene is one of the newer entrants into the carpet industry and is finding favour because of its good resistance to abrasion and soiling.

Vegetable fibres are used to make low-cost looped pile carpets. The fibres employed are either jute or sisal and occasionally cotton, although the latter makes a poor pile and dirties easily.

Mineral fibres are only used in conjunction with other fibres and are added to provide electrical conductivity to dissipate static charges which, because of the dry atmospheres now achieved in many buildings, are an ever present problem.

Not only does the material used for the pile affect the character of a carpet, so also does the way the carpet is made up. The three main methods now used are weaving, tufting and needle punching.

All carpets comprise the base or backing and the pile. With a woven carpet the pile is looped or run in the backing as the latter is woven. The traditional weaves for achieving this are Wilton and Axminster. The Wilton weave incorporates more of the pile fibre in the backing than the Axminster weave and thus produces a heavier, better-quality carpet which is often used in hotels, restaurants and similar buildings. Cord carpet is also produced using the Wilton weave but is left with the pile uncut.

Tufted carpets are made by stitching yarn onto jute or polypropylene backing and securing it with an applied rubber backing. Production is much faster than weaving and the quantity of material used is smaller, consequently these are less expensive carpets even when the same pile fibre is used.

Needlepunch or fibre-bonded carpets are made by laying out a mat of fibres and punching or pressing them into a dense felt. This is then treated with a bonding agent and sometimes has a foam backing applied.

Other methods of producing carpets, less commonly employed, are flocking and bonding. Flocked carpets are made by blowing electrically charged cut fibres onto a base coated with adhesive. The electrical charge makes the fibres stand on end until the adhesive has set. Bonded carpets are also made with adhesive which is spread onto a woven jute backing and into it are pressed V-shaped tufts of pile fibre.

Many manufacturers are still producing and marketing carpets in imperial widths of 27 in, 36 in, 54 in, 7 ft 6 in, 9 ft 0 in, 10 ft 6 in, 12 ft, 13 ft 6 in and 15 ft 0 in. The term 'broadloom' is applied to any carpet over 54 in wide. Metric sizes are 1, 2, 3, and 4 m wide. All carpet manufacture is a continuous process, so there is no actual limit to the possible length of the roll, but in practice this is kept to 25 m to avoid problems of handling.

Unless the carpet, whatever its construction or fibre, has its own high-density foam backing it should be provided with an underlay to prolong its life. If it has an integral foam backing or underlay the introduction of another can actually shorten its life. Foam backed carpets are stuck directly to the sub-floor using a 'peel up' technique involving the use of a 'release sealer' applied to the floor in advance of the adhesive to allow the carpet to be taken up without loosing its foam backing. Where the carpet has an underlay, this is laid first, inside carpet gripper battens nailed to the floor round the perimeter of the carpeted area. Gripper battens are timber or metal strips fitted with rows of small hooks or inclined spikes. The battens are fixed so that the spikes point towards the wall. The carpet is placed over the underlay and cut slightly undersize for the room. The cut edges are then treated to prevent fraying, if necessary, and the carpet stretched and secured to the gripper batten. Underlays are either of foam rubber, felt or a combination of the two materials. Rubber offers a longer-lasting resilience and better impact sound absorption, but felt is cheaper and has a better airborne sound absorption. Where a carpet has been woven in the narrower widths and the seams sewn together (a practice often followed with the better qualities) it must be laid on a felt underlay into which the seams can bed.

19.4 Tile flooring

The advantage of tile flooring over sheet flooring is ease of laying and economy of materials. The economy arises because the quantity of tiles used more nearly equates to the actual floor area, whereas sheet material must be obtained in a size out of which the area to be covered can be cut. Where the room size is very different to the sheet size, or there are a lot of fixtures round which it must be trimmed, wastage of sheet material can be high. It is also possible, by careful selection and laying of tiles, to produce a design in the floor finish which is unique to the room.

All the materials dealt with in the preceding section are also produced in the tile form. In the case of vinyl the tile material is no different to that used for the sheet flooring. Carpet tiles are much the same as carpet on the roll, but with additional treatment to prevent fraying and to improve their lay. Certain carpet tiles of a felted nature and some using animal fibre are only produced in this form. Rubber tiles are either cut from the sheet flooring as produced or are moulded with a ribbed or studded surface.

The other forms of tile now available are thermoplastic, vinyl asbestos, cork, clay, concrete and terrazzo.

Thermoplastic tiles soften when warmed and are made from thermoplastic binders, plasticisers, short-fibre asbestos, powdered mineral fillers and pigments which are heated and pressed into a sheet and cut into tiles. The material cannot be used in the sheet form because it is brittle at normal temperatures. The colours available tend to be dark, because the binders most commonly used are pitch or bitumen. It is a cheap, hard-wearing flooring produced in tile sizes of 300 × 300 mm and 250 × 250 mm in thicknesses of 2.5 and 3 mm.

Vinyl asbestos tiles are similar to thermoplastic but using a polyvinyl chloride binder. The material is not flexible so is only available in tile form. It lies between thermoplastic tiling and vinyl tiling in cost, the colours available are better than thermoplastic because the binder is white and the sizes produced are 300 × 300 mm and 225 × 225 mm in thicknesses of 1.6, 2.0, 2.5 and 3.2 mm.

Cork tiles are made from the bark of the cork-oak tree which grows in Southern Europe, North Africa and the East. The cork is reduced to granules and then heated and pressed. No binders are used since the natural gum in the cork holds the particles together after the heat treatment. In many makes, a hot wax surface finish is applied to fill the large pores in the open texture or a thin layer of transparent plastic is bonded to the surface. Most manufacturers produce tiles 305 mm square in thicknesses from 3 to 12 mm and in various densities.

19.4.1 Clay tiling

There are three main types of clay floor tile: quarry tiles, vitrified or semi-vitrified tiles and ceramics. They are all made from naturally occurring clay selected for a particular colour and refined in varying degrees.

Quarry tiles are the coarsest and heaviest; the clay used is the least refined, the colours available are restricted to about four. They are hard impervious tiles varying in thickness from 1 in (25.4 mm) to $1\frac{1}{2}$ in (38.1 mm) in a wide variety of sizes but usually in the 250 to 300 mm square range and are usually laid in mortar with wide joints to allow for the irregularities which occur in the tile shape.

Vitrified and semi-vitrified tiles, or tesselated tiles as they are sometimes called, are produced from raw materials which are more refined than that used for quarries. Various oxides are added to give a greater colour range. They are produced in a great range of sizes and many different shapes.

Ceramic tiles are based on highly refined clay as a raw material, to which is added glass, quartz, basalt and other similar materials in a carefully proportioned mix which, when fired at 1200 °C will produce the desired result. For heavy use such as a laboratory floor, fully vitrified unglazed tiles are best but where the use is lighter, such as in a domestic bathroom, the glazed type is adequate and offers a much larger range of colours and patterns.

While tiles can still be laid in the traditional manner with a sand/cement mortar bed, the more common practice now is to use one of the cement-based adhesives. The use of these generally requires an accurate surface on the sub-floor since they provide only a thin bed which cannot correct any irregularities. There are several types of adhesive manufactured for differing locations and use of floor and reference should be made to the specifications published by the British Ceramic Tile Council.

After laying, the joints between the tiles must be grouted; if they are laid on a cement mortar base the grouting should be a mix of 1:1 Portland cement and fine sand (white cement is often used to produce a better joint colour); if the tiles are laid in adhesive a compressible grout recommended by the adhesive manufacturer (and usually made by him) must be used. Tiles in cement mortar are firmly held and so a hard grouting is suitable but adhesive allows a slight movement to accommodate any dimensional changes in the tiles or floor. A non-compressible grout can restrict this movement, causing the tiles to arch and lift off the floor.

19.4.2 Concrete and terrazzo tiles

Concrete tiles are made from normal concrete materials possibly with additives to make them water repellant and to give colour. They are hydraulically pressed to make them hard and are produced in a range of sizes from $150 \times 150 \times 15$ mm up to $500 \times 500 \times 40$ mm.

Terrazzo tiles are of similar construction to concrete tiles, but have a facing of good quality marble chips in white or coloured cement which is ground and polished to a fine finish. Tile sizes are the same as for concrete tiles.

19.5 Jointless flooring

There are a number of proprietary products which can be applied to a sand/cement screed by brush or spray to produce a finished surface, but most jointless floors are trowel applied. The cheapest trowelled finish is granolithic, which really is a better quality cement screed using a carefully selected and graded aggregate of a suitable hardness – often granite chippings and granite dust – laid and finished in the same way as a normal screed.

A rather better version of granolithic is terrazzo, in which the aggregate is marble and the cement either white or coloured. After laying, the surface is ground and polished to reveal the beauty of the natural stone. The size of panels should be kept small and separated with ebonite strips, to avoid cracks disfiguring the floor. This method produces a beautiful but hard floor which, because of its high polish, can become slippery if wet.

Other cement-based floorings are cement rubber-latex and cement resin emulsion finishes. These consist of a mixture of Portland cement with aggregate such as cork, sand, sawdust or wood flour or stone or marble chips and a filler, gauged with an emulsion of rubber-latex or resin. It is intended that it should be laid on a concrete floor but it will adhere to a wooden floor as well. However, reinforcement of wire mesh should be used where it is laid over timber.

A non-cementitious flooring which has been in use for many years is magnesite or composition flooring. This consists of a combination of calcined magnesite, wood flour, sawdust, ground silica, talc and powdered asbestos, compounded with a solution of magnesium chloride. While it is resistant to alkalis, organic solvents, salt and sulphates, oils, greases and fats, it has little tolerance to continued wetting.

After laying, the floor is dressed with equal parts of turpentine and boiled linseed oil followed by a sealant based on acrylic, polyurethane or epoxy resins.

Mastic asphalt with various aggregates of differing degrees of hardness produces a quiet, hygienic, jointless floor favoured in certain situations. Its colour range is limited to black, dark red or dark brown and dark grey or green, but it is a dust-free surface which is extremely wear resistant to wheeled traffic, non-slip, non-inflammable and water resistant. However, asphalt is not resistant to milk, oils, fats or grease and should not be used in areas where these are likely to be regularly spilt; it can also be softened by the solvents used in some floor polishes.

Chapter 20

Suspended ceilings

20.1 General considerations

With the exception of plaster directly applied to the soffit of a solid
concrete floor slab, all ceilings are suspended to some degree. The
term 'suspended ceiling' is intended in this chapter to refer to an
arrangement where the material with which the ceiling is finished is
attached to a framework which itself is hung at a distance from the
soffit of a structural floor or roof. The void thus created may be
used for services or the reverse situation may apply where, because
services have to be run below the floor soffit, a suspended ceiling is
provided to conceal them.

The first point to be considered is whether a suspended ceiling
system is required. There are many advantages to be gained by
having a continuous void above all the rooms in a building, but this
can only be achieved at an additional cost and unless the building
use is such that a demand is created for services or provisions which
can only be satisfied by a suspended ceiling, then this additional cost
cannot be justified. It is not simply in the provision of the ceiling
that the extra costs arise; the additional depth from floor surface to
ceiling surface must be accommodated in the building, which means
greater floor to floor heights, a taller building and longer vertical
elements such as rising mains, drainage stacks, staircases, lifts, etc.
For this reason, even when the decision is in favour of a suspended
ceiling, the distance between the floor soffit and the ceiling is kept as
short as possible.

20.1.1 Access

As already mentioned, one of the reasons for choosing a suspended ceiling is the concealment of services. Where this is done, it is also necessary to provide means of access to those services for maintenance and possible alterations.

The simplest and cheapest way of achieving this is not really a ceil..g at all but a grid of laths fixed below the service runs. The laths are spaced at 600 mm centres each way, and are painted white or cream. They often support modular lighting fittings, which are made 600 mm square. Everything above the level of the grid is painted dark brown or dark green. The design relies on the colour contrast between the grid and the dark paintwork above, plus carefully positioned light fittings, to create the illusion of a ceiling.

A suspended ceiling comprising a framework into which panels are laid or slotted also affords ready access to the ceiling void, since a panel or group of panels can be lifted out whenever the services require attention. The total accessibility afforded by a panel ceiling or the grid arrangement is not usually necessary; there are specific points in any system of pipes, ducts or cables, to which access is required and removable panels below these points should provide the engineer with all his needs.

20.1.2 Acoustic control

Besides the concealment of services a suspended ceiling may be installed for purposes of noise control. Ceilings of perforated or slotted units backed by glass fibre or mineral wool provide good sound absorption and thus are very popular in large or open plan offices, where the background noise level must be kept low. If the slots or holes are very small, care must be exercised in their maintenance, since cleaning or painting can fill these openings and the sound absorbency of the system is lost. Few suspended ceilings offer much resistance to sound transmission and therefore the partitions separating areas where sound insulation is important must be carried up to the structural floor to close off the sound transmission route through the ceiling void.

20.1.3 Fire protection

Fire resistance is important but most types of suspended ceiling can achieve a 2 hour resistance and a Class 1 spread-of-flame rating. Where the ceiling is required to contribute to the overall fire resistance of the floor, this spread-of-flame rating is adequate for buildings less than 15 m high where the floor is not a compartment floor or a compartment floor of less than 1 hour fire resistance. In any other case the spread-of-flame rating must be Class 0, as set out in Approved Document B2 to the Building Regulations.

The resistance of the ceiling to the passage of fire is one aspect of fire protection. The other is the danger of a fire, having broken through the ceiling, rapidly spreading through the ceiling void, to carry the conflagration to all parts of the building. To prevent this, fire stops, specified in Approved Document B2/3/4, are fixed to close the void along the lines of any fire resisting partitions, and at 20 m intervals if the spread-of-flame rating of the surface exposed to the void is Class 0, or 1 or 10 m intervals if the rating is any other class.

20.2 Types of suspended ceiling

From the constructional point of view there are four types of suspended ceiling: jointless, panel, strip and open. Within each of these groups are ceiling systems which can meet most of the performance requirements which may be specified for any room.

20.2.1 Jointless ceilings

Most jointless ceilings are of plasterboard, plaster lath or metal lath with a plaster finish of either one or two coats. They are fixed to a supporting framework hung from the floor soffit above which can be either a purpose made timber structure or, as is more commonly the case, a grid of small mild steel channels suspended on wire hangers or mild steel straps (see Fig. 20.1): A useful publication to consult when a metal framework is to be used is issued by the Metal Fixings Association for ceiling systems under the title *Code of Practice and Manual of Lightweight Metal Fixing Systems for Building Linings and Ceilings*.

Access to the ceiling void is restricted to points where traps are provided. Since these tend to disfigure the appearance, jointless plaster ceilings are not a first choice where a complex service installation needs to be concealed. Their use is mainly confined to rooms or areas where it is desired to reduce the height.

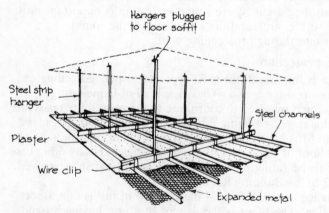

Fig. 20.1 Suspended jointless ceiling

20.2.2 Panel ceilings

Panel ceilings are by far the most common form of suspended ceiling and consist of infill panels of a wide range of materials and finishes fitted to a metal suspension framework.

The metal framework system usually consists of hangers of straps, angles or, more commonly, rods, secured to the floor structure above. The hangers connect to bearers via an arrangement that affords levelling adjustment and the bearers support runners fixed at right angles to them. Subsidiary cross members, called noggins, are clipped between the runners and the infill panels are supported by the runners and the noggins. The bearers and hangers are usually of steel and the runners and noggins of aluminium (see Fig. 20.2).

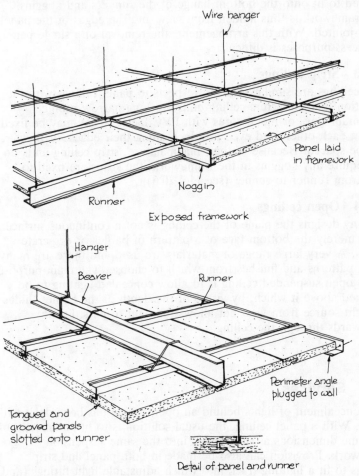

Wire hanger

Panel laid in framework

Noggin

Runner

Exposed framework

Hanger

Bearer

Runner

Perimeter angle plugged to wall

Tongued and grooved panels slotted onto runner

Detail of panel and runner

Concealed framework

Fig. 20.2 Suspended panel ceilings

The panels are produced in a wide range of materials either as flat sheets, with various surface textures and colours, in some cases perforated and with sound absorbing backing, or in a shaped or profiled form to give the ceiling a deeply sculpted effect. Fixing of the tiles to the framework is usually either 'lay-in' or 'concealed' (see Fig. 20.2). With the lay-in method, the runners and noggins are T section members and the panels simply rest on the flange of the T, which remains exposed.

In some systems the panels are retained by a spring clip. This method affords easy means of access because a panel can be readily lifted out anywhere over the area of the ceiling. The concealed method generally uses a Z or channel section and the panels are grooved to fit onto the bottom flange of the runners and noggins. The framework is thus hidden from view and the edges of the panels are V-jointed. With this arrangement, the removal of a single panel for access purposes is difficult.

20.2.3 Strip ceilings

Strip ceilings are similar in many respects to panel ceilings with the exception that the actual ceiling units are long narrow profiled aluminium trays or metal cores with PVC facing. They may be fixed to abut each other, and even interlock along their edges, or they may be spaced. Because of their linear nature, strip ceiling units do not require any noggins in the framework, the strips are fixed to span from runner to runner (see Fig. 20.3).

20.2.4 Open ceilings

In many designs the plane of the ceiling is not a continuous surface but is merely the bottom face of a pattern of baffles or egg-crate panels. A very large range of materials are used and there are many types, patterns and finishes from which to choose. The main purpose of the open suspended ceiling is to allow concealed lighting to be arranged above it which, by virtue of the depth of the baffles, hides the light source from view but allows the light to pass directly downwards through the ceiling.

20.3 Services

The concealment of lights behind an open ceiling has been dealt with above. With a panel ceiling, the usual solution is to fit luminaires, of the same dimensions as the panels, into the same suspension framework. Provision can also be made in both panel and strip ceilings to fit a lighting track into which adjustable light fittings (and also speaker units) can be plugged, as required. With jointless ceilings, the light fittings would probably be of the type fixed to non-

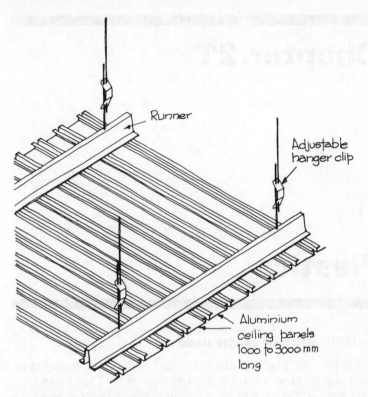

Runner

Adjustable
hanger clip

Aluminium
ceiling panels
1000 to 3000 mm
long

Fig. 20.3 Suspended strip ceiling

suspended ceilings, but in this case fitted to the ceiling surface and
fixed to the support framework behind.

Ventilation is the other service which makes use of the
advantages of a suspended ceiling. Not only is a space available in
which to run the necessary ducting, but it is also possible to
incorporate the extract units into either a panel or strip ceiling so
that they follow the ceiling pattern, or they can be mounted behind
an open ceiling and thus are completely concealed.

Chapter 21

Plastics

21.1 Definition of terms used

Many materials are plastic in the sense that they can be moulded in
given conditions, or at a certain stage in their lives. Thus by this
definition steel, concrete and glass are all plastic. Plaster is also
plastic and, indeed, both words are derived from the same Greek
root 'plastikos', meaning mouldable.

Plastics is now used as a singular noun relating to mainly organic
petrochemical materials which are plastic when heated. The
development of the plastics industry has brought with it a number of
terms which are used regularly and which are briefly defined below.

21.1.1 Saturation and unsaturation

The simplest substances we know, i.e. they cannot be split into any
other component substances, are called elements and the smallest
particle of an element is an atom. Most substances are compounds of
the elements, produced by atoms linking together into molecules.
This linking is achieved by a chemical bond between the atoms. An
organic compound (one based on a carbon atom linked with some
other atoms) in which each carbon atom is linked by a single bond
to four other atoms, is known as a saturated compound. If, however,
the carbon is linked by a double or triple bond to some other atom,
then the compound is unsaturated and has a certain chemical
instability. Ethylene is a typical unsaturated compound. Breakage of
one of the double bonds creates a situation where another similarly

ruptured molecule can join to the first to form a chain. If this happens to ethylene, one of the chemicals in the paraffin group is formed. The process can be allowed to repeat to form a long-chain molecule such as is found in polythene, or more correctly, polyethylene. This process is known as polymerisation, and the end result is called a polymer.

21.1.2 Polymerisation and polymers

When the molecules of the monomer (the small unit forming the link in the chain, e.g. ethylene in the example above) react in the manner described in the previous section to build a long-chain molecule, it is referred to as an additive polymerisation and there can be as many as 20 000 links in a single chain.

Condensation polymerisation is the term used to describe a reaction involving two different kinds of monomers, the product of which is either a long-chain molecule or a three-dimensional cross-linked structure. In either case, molecules of such simple substances as water or hydrogen chloride are eliminated, which fact distinguishes condensation polymerisation from additive polymerisation.

When two or more different monomers are polymerised the product is known as a copolymer. This product often possesses merits more desirable than those of the polymers formed separately from each monomer – in the same way that an alloy is better than the constituent metals.

Polymers can be classified as:

Natural: rubber, proteins, cellulose, etc.;
Semi-synthetic: cellulose acetate, nitrocellulose (celluloid), etc. (both made from cotton);
Synthetic: polymethyl methacrylate (Perspex), phenol formaldehyde (Bakelite), polystyrene, nylon, etc. (Perspex and Bakelite are trade names belonging to Imperial Chemical Industries and Bakelite Xylanite respectively).

Within these classifications there are three polymer structures: rubbers, thermoplastics and thermosetting plastics.

21.1.3 Rubber

The term rubber refers to a polymer with elastomeric properties. Its molecular chains are coiled and arranged in a random manner, which makes it an amorphous material (non-crystalline) and the electrostatic forces linking the molecular chains (the van der Waals forces) are small. When tension is applied, the molecular chains straighten out and the rubber stretches. The structure is such that on release of the tension the chains can recoil with little interference from neighbouring chains and the rubber returns to its original size.

21.1.4 Thermoplastics

The structure of thermoplastics is partly amorphous – with its molecular chains entangled – and partly crystalline – with its molecular chains straight. Where a chain is straight no lengthening can occur under conditions of tension, thus thermoplastics are less elastic than the rubbers.

The van der Waals forces in thermoplastics are small and easily overcome by a small rise in temperature. When this occurs, the distances between the chains can change and the material softens. Once the temperature returns to normal, the van der Waals forces reassert themselves and the material sets in its new shape. This can be repeated indefinitely to give thermoplastics their characteristic of being able to be heated and moulded under pressure.

21.1.5 Thermosetting plastics

In this polymer structure the cross linkage between the molecular chains is strong and permanent, producing a three-dimensional molecular structure which is rigid and inelastic. Heat is required to produce the structure initially but subsequent application of heat will not soften it. The material is hard, brittle and relatively scratch resistant in its moulded form and insoluble in organic solvents such as benzene or acetone.

21.1.6 Plasticisers

Plasticisers are materials added to polymers to lower their softening temperature and make them less brittle. Polyvinyl chloride is a plastic commonly used in buildings in its natural, or unplasticised form (UPVC). It is rigid or stiff enough to use for rainwater goods, water mains, drain pipes, electrical conduits and accessories, and for roofing sheets (where its self-extinguishing properties are welcome). If a plasticiser is added, the material (PVC) becomes flexible and can be extruded, injection moulded, calendered or blown into film (see next section on processing) and thus used for floor coverings, sarking, water stops, cable insulation, washable wallpaper, etc.

21.1.7 Fillers

The addition of inert solids to plastics is done either to lower the cost by increasing the bulk without changing the characteristics, or to modify the properties of the material. Wood floor, alpha cellulose and cotton, glass or synthetic fibres reduce brittleness and increase impact strength; asbestos fibres, mineral powders and mica flakes improve the heat resistance or the chemical resistance or both.

21.2 Processing plastics

Plastic materials leave the manufacturer in the form of fine powders, emulsions, resins, viscous fluids, pellets, granules, cubes or sheets, and these must be processed to render them into workable material from which the articles we use in the building industry can be made.

The main processes employed can produce either a continuous length of the required product, or individual units manufactured either by moulding plastic powder or granules, or by forming them from plastic sheets.

There are many methods and variations of each process but the main ones are illustrated in Figs 21.1 and 21.2, and briefly described below.

Continuous production is achieved by extruding, film blowing, calendering and paste spreading; moulding methods are mainly compression moulding, injection moulding, blow moulding and rotational moulding; thermo-forming from sheet material is done by pressing, vacuum forming, or laminating.

21.2.1 Extruding

This gives the most economical output of all plastic shaping methods. The products must be simple in shape and, being a continuous process, only two dimensions are defined in production. Thus pipe, channels, gutters, curtain rails, mouldings, trims, electric cables and films are typical end products.

The process is to feed plastic powder or granules onto a screw rotating inside a heated cylinder. The rotation of the screw forces the plastic forward and the heat converts it to a 'melt' of the right consistency to extrude through the die in the required profile. It is then quickly cooled, usually in a cooling bath. Foamed plastics can be extruded by the same process.

21.2.2 Film blowing

A major demand in the building industry is for plastic film and consequently the method of producing this is one of the more important processes. It is simply achieved by extruding a tube and then rapidly expanding it by blowing air into it. The extent of expansion and hence the film thickness is controlled by the air pressure and a cooling ring which sets the plastic film at the required size. At the top of the blowing chamber are a pair of nip rollers, which close the top of the tube and also extend the film longitudinally, by running at a linear speed which is faster than the extrusion rate. From the rollers the film tube passes to a wind-up unit.

21.2.3 Calendering

Calendering was originally developed by the rubber industry to produce sheeting and has been adopted for the same purpose by plastics manufacturers, especially for plasticized PVC sheets.

The process is that a molten polymer is fed into a pair of rollers

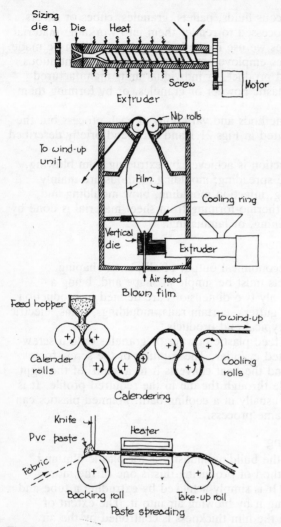

Fig. 21.1 Continuous plastics processing

which are the first of a train of such rollers. The first two pairs are
the calender rolls; they are heated and the space between them, the
'nip', controls the sheet thickness. Following the calender rolls are
the cooling rolls, whose name indicates their function.

Very high output can be maintained by the process for a low
energy consumption, when compared to extruding. It is also possible
to use the method for laminating plastics film onto other materials
such as textiles, paper and metal foil.

21.2.4 Paste spreading

In this process a paste of plastic, usually PVC, is poured onto a moving backing and spread to a uniform thickness between a backing roll and a knife. The coated fabric is then heated to cure the plastic and wound into rolls of specified length. Vinyl faced fabrics for upholstery and clothing are produced in this manner as also is washable wallpaper.

21.2.5 Compression moulding

This method generally uses a thermosetting plastic in powder form, or a cube of predetermined size to suit the article to be produced. The process is to place a measured quantity of powder or a cube into the bottom half of the mould, close down the upper half while at the same time applying heat at between 150 °C and 200 °C. The combination of pressure and heat causes the plastic to flow into all the cavities of the mould. The conditions are held until the material has cured and all the molecules have firmly cross linked to stabilise the moulded shape, and then the mould is opened. WC seats and door handles are produced in this way.

21.2.6 Injection moulding

Thermoplastic materials are used for injection moulding. The raw material is fed onto a screw rotating inside a heated cylinder, very similar to an extrusion cylinder, in which the plastic is melted and transported forward but, instead of the screw extruding the material, the collecting 'melt' at the nozzle forces the screw backwards until a specified quantity has collected. The screw is then pushed forward by hydraulic pressure, thus acting as a plunger to inject the melt into the mould. Whilst the melt is cooling in the mould, the screw is transporting and preparing another 'shot' of melt to be injected when the cooled moulding is ejected. It is an automatic process, well suited to large production runs, and the mouldings can vary in size from tiny gearwheels to small boats and can be open, like cisterns, manholes, buckets or bowls, or complex like telephones.

21.2.7 Blow moulding

In this process a predetermined length of thermoplastic tube, called a parison, is nipped between the top and bottom of a mould as it closes. This welds one end of the tube together and seals the other round a metal tube. Air blown in through the metal tube expands the parison until it takes up the shape of the mould. Pressure is maintained until the component has cooled, when it is released from the mould.

Blow moulding is used to make closed containers of all sorts – small hollow components, 20 mm in diameter or square, bottles of various sizes and containers up to 1.5 m^2 capacity. Cold water cisterns can

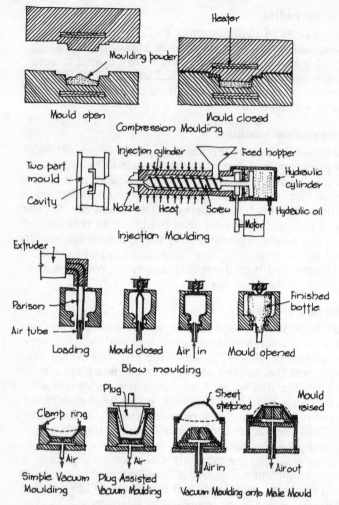

Fig. 21.2 Plastic moulding processes

be made in this way by blow moulding an enclosed 'box' and cutting it in half to form two cisterns.

21.2.8 Rotational moulding

Cold water cisterns can also be formed by rotational moulding, as also can dustbins, boat hulls and any hollow component of simple shape. As the name implies, the mould is rotated about two axes while the plastic powder, which has been placed inside, coats itself over the inner surface of the mould. The mould is heated while it is rotating to melt the powder into a uniform continuous layer. When all the powder has coated the inside of the mould it is cooled, and the component removed.

21.2.9 Pressing

Pressing is the simplest of the thermo-forming methods and consists of clamping a heated sheet of thermoplastic across a die plate into which has been worked the outside contours of the component. The sheet is pressed into the die plate by a former, shaped to the inside contours of the component. When it has cooled the pressure is released and the component removed.

21.2.10 Vacuum forming

This method also uses a thermoplastic sheet which is heated and clamped to a die plate but in this case the mould is penetrated by small tubes connected to an air pump. When the air is extracted the sheet is sucked down onto the mould and assumes its shape. The procedure may be assisted by stretching the sheet first, either by the use of a plug, rather like a former, or by blowing in compressed air, which expands the sheet upwards, and then drawing it down onto a male mould.

21.2.11 Laminating

Many worktops and surfaces in present buildings are faced with laminated plastics. These are produced from layers of thermosetting, resin-impregnated paper or cloth, pressed and heated to form a single sheet. Their most familiar use is as a hardwearing decorative sheet, which is generally built up from layers of kraft paper impregnated with phenol formaldehyde. The top layer of paper carries the decoration and is covered with a veneer of clear melamine formaldehyde. Other products from this source are gearwheels and other mechanical parts, and the circuit boards used in much of our present electronic equipment.

BS EN 438: 1991 specifies that high pressure laminates (HPL) should be classified by either a type and index number system indicating resistance to wear, impact and scratching (in each case on a scale of 1 to 4) or an alphabetical classification system based on grade, class and type. There are three types, S (standard), F (flame retardant) and P (post formable); three grades, H (for horizontal worktop use), V (for vertical panels, doors, etc.) and C (compact self-supporting panels 5mm or more in thickness) and two classes, G (general purpose) and D (heavy duty).

Either of the systems can be used but should be prefixed by HPL, to indicate the material and the BS number, EN 438. Thus, a kitchen worktop laminate may be: HPL-EN 438–P333 (Postformable with high resistance to wear, impact and scratching) or: HPL-EN 438–HGP (horizontal, general purpose, postformable).

21.3 Uses of plastics

Table 21.1 below sets out the main plastics encountered in the building industry and a typical range of uses made of them.

Table 21.1 Uses of plastics materials

Material	Abbreviation	Uses	Comments
THERMOPLASTICS			
Polyethylene	PE	Cold water cisterns and floats. Cold water pipes. Waste pipes. Damp-proof coursing and membranes. Transparent sheet for general purposes.	Usually called polythene. Short life unless carbon black is added to the mix.
Polyvinyl chloride	UPVC	Drain and sewer pipes. Water mains. Rainwater goods. Translucent profiled roofing sheets. Safety glazing. Electrical conduits and fittings. Window frames Insulation (expanded form).	Rigid natural form without added plasticiser
	PVC	Floor coverings. Sarking. Water stops and seals Handrails Electrical cable insulation Flat roofing (single layer). Insulation (expanded form).	Plasticiser added to give flexibility.
Polyvinyl acetate	PVAC	Joinery adhesive. Paints. Bonding agents.	Low softening point.
Polypropylene	PP	Pitch fibre pipe fittings. Waste pipes. Water tanks. WC cisterns and seats. Chair shells. Carpets.	High softening temperature. Hard wearing.

Material	Abbreviation	Uses	Comments
Polymethyl methacrylate	PMMA	Transparent roofing. Roof lights. Light fittings and signs. Baths and basins. Shop fascias. Lettering.	Better known as Perspex.
Polystyrene	PS	Wall and ceiling tiles. WC cisterns. Light fittings. Insulation boards and beads (expanded form).	Brittle. High impact and medium impact grades available.
Polytetrafluorethylene	PTFE	Sliding bearings in heavy structures. Pipe thread sealing tape. Non-stick linings.	Expensive. High resistance to heat. Very low coefficient of friction.
Acrylonitrile butadiene	ABS	Garage doors. Machine housings. Waste pipes. Inspection chambers. Industrial pressure pipes.	Tough, strong, high impact strength.
Cellulose nitrate	CN	Paint.	Originally called Celluloid. Highly inflammable.
Cellulose acetate	CA	Light fittings. Door handles.	Similar to Celluloid but not so inflammable.
Casein	CS	Adhesive.	Made from milk whey.

Material	Abbreviation	Uses	Comments
Nylon (aliphatic polyamine)		Nuts and bolts. Castors. Curtain rails. Sliding door fittings. Ball valves. Door and window furniture. Hinges.	Many forms, Nylon 6, Nylon 11 and Nylon 66 most important for building.
Polyacetate	POM	Plumbing fittings.	Similar characteristics to metal.
Polycarbonate	PC	Vandal resistant glazing. Bullet proof glazing.	Dense, hard, tough and with high tensile strength.

THERMOSETTING PLASTICS

Material	Abbreviation	Uses	Comments
Phenol-formaldehyde	PF	Electrical fittings. Door furniture. WC seats. Paint. Adhesive.	Better known as Bakelite.
Urea-formaldehyde	UF	Electrical fittings. Caps and lids. WC seats. Cavity wall insulation (UF foam).	High electrical resistance. Wide colour range.
Melamine-formaldehyde	MF	Surfacing decorative laminates. Door handles. Table and kitchen ware.	Hard. Resistant to heat and staining.
Resorcinol-formaldehyde	RF	Timber adhesive.	

Materials	Abbrev-iations	Uses	Comments
Polyester resin	UP	Solar control film. Floor coverings.	Wide range of properties.
Glass fibre reinforced polyester resin	GRP	Roof sheet. Roof lights. Cold water cisterns. Hot water cylinders. Cladding panels. Inspection chambers. Cesspools. Shower cubicles. Pipes. Trims. Channel and tube. Sections for general purposes.	Glass fibres most common but others are used. High strength/weight ratio.
Polyurethane	PU	Paint. Clear finishes. Sealants. Insulation (rigid or flexible foams).	Covers a multiplicity of plastics derived from polyisocyanates
Epoxy resin	EP	Floor coating. Road surfacing. Adhesive. Clear finish. Electronic components.	Tough, stable, good adherence. Good electrical properties and chemical resistance.
Silicone resin	SI	Waterproofer (paint-on or inject). Floor polish.	Water repellent.

Part V

External works

Chapter 22

Road construction

22.1 Rigid pavements

In the civil engineer's vocabulary, the word pavement is used to describe the part of the road along which vehicles travel and, more specifically, that part of the road above the sub-grade, or upper part of the soil below the pavements. The part along which pedestrians walk is referred to as the footway.

Chapter 21 of *Advanced Building Construction*, Vol. 1 deals in some detail with the terminology and design of rigid pavements, particularly the joints used, and this chapter will deal with the other aspects of road construction: edge details, block paving, drainage and setting out.

22.2 Road edge details

The function of the kerb provided along the edge of a road is mainly to strengthen the carriageway edge and to prevent lateral displacement of the pavement by traffic. It will also control surface water run-off and can prevent both the traffic leaving the carriageway at hazardous joints and vegetation encroaching onto the carriageway. Most kerbs give a firm clean line which defines the edge of the pavement and, when laid in advance of the pavement, can be used as a level guide for the various pavement layers. On major road schemes, the pavement laying machines are guided by a line stretched at the edge of the formation.

The edges of estate roads, and many of the more important roads in this country, are finished with a kerb; motorways and many trunk roads, however, are finished by a hard shoulder, and present practice is to finish all new roads in rural areas with a hard strip, which is similar to a hard shoulder but only 1 m wide.

22.2.1 Kerbs

Figure 22.1 shows the profiles of the kerbs in use at the present time and comprise raised kerbs, flush kerbs and kerb/channel combinations.

The raised kerb is frequently used and of the types available the half-batter kerb is the most common, especially where a footway adjoins the carriageway. This type has the advantage of protecting the footway to some extent but yet allowing vehicles to drive over it if necessity directs. Where protection of the footway is of primary importance a vertical kerb is used which offers more resistance to a wheel crossing it. If vehicles are to be allowed to leave the carriageway in an emergency, splayed kerbs are laid. The barrier kerb is less common on our roads, and is provided to confine vehicles to the carriageway at sharp curves, high embankments, etc., where the requirement is to slow the vehicle under control, preventing it from going over the edge of the road or crossing into another traffic lane.

Where there is no footway to protect, or where vehicles are to be allowed to leave the carriageway, flush kerbs are laid. The simplest of these is the plain flush kerb whose purpose is merely to strengthen and contain the edge of the pavement. A serrated flush kerb serves the same purpose as a plain flush kerb, but also gives warning to a driver when his vehicle is leaving the carriageway. Dished channel kerbs are provided where surface water control is also required.

If, as is common practice, the kerb is laid in advance of the pavement, a vertical joint between the two occurs at the point where surface water collects. Penetration by this water can soften the sub-grade under the pavement edges and, if it freezes, it can displace the kerbs. To overcome this, the combined kerb and channel has been developed which provides a raised kerb, either vertical, half battered or splayed, and a flat channel section along which surface water is directed. A variation of this is the continental kerb, which combines a dished channel with a rounded mountable kerb.

22.2.2 Hard shoulders

Motorways are provided with hard shoulders, normally 3.3 m wide, alongside each carriageway to allow vehicles to leave the traffic lane in the event of a breakdown. They also provide an access route for emergency services if an accident occurs, or an extra traffic lane during maintenance.

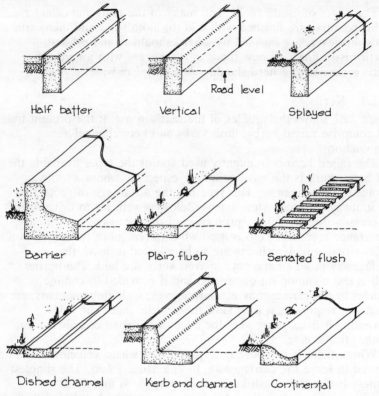

Half batter   Vertical   Splayed

Barrier   Plain flush   Serrated flush

Dished channel   Kerb and channel   Continental

Fig. 22.1   Kerb profiles

The construction of a hard shoulder does not need to equal that of the main carriageway since its use is less frequent and traffic speed is also reduced, but it must be strong enough to support the heaviest vehicle which may use the motorway, and this support must allow for the vehicle to be jacked at one point. It must also permit the free drainage of surface water off the carriageway.

The hard shoulder provides the carriageway with edge strengthening and lateral displacement control normally provided by a kerb. In these circumstances, the kerb is moved out to the edge of the hard shoulder, where as well as confining the hard shoulder pavement, it also gives an edge definition which is visible in snow.

22.2.3   Hard strips

A hard strip, as with a hard shoulder, performs the function of a kerb to the carriageway but, usually, is not itself provided with a kerb since it is generally a road in a rural area which is being constructed, and surface water can be allowed simply to run off the edge (possibly into a French drain – see section 20.5).

22.3 Kerb design and construction

The correct choice of cross-section of a kerb is important and must relate to the type of road and the kerb function (edge definition, surface water control, protection of footway, etc.). It is also necessary to select the 'upstand' or the height of the top of the kerb above channel level with care.

In deciding the upstand of a non-mountable kerb, the convenience of pedestrians, especially women with prams, as well as the protection of the footway must be considered. Kerbs which are excessively high can cause trouble for pedestrians and can also effectively reduce the width of the carriageway due to 'kerb-shyness' on the part of the motorist fearful of damage to his car wings, cills and hubs. The *Report of the Committee on Highway Maintenance* under the chairmanship of A. H. Marshall suggested, in 1970, that a large upstand of 175 mm should be provided to allow for resurfacing. It is now considered that this is too great since resurfacing of residential streets is likely to be an infrequent application of a thin coat, and that 100 to 150 m maximum is better. An exception to this is in heavily trafficked areas where an upstand of 175 mm is preferable, to reduce the risk of large commercial vehicles overriding the footways.

The materials used for kerbs are concrete, natural stone and asphalt. Concrete kerbs are either pre-cast, or run *in situ*; stone kerbs are composed of 'setts', usually of granite or whinstone. Asphalt kerbs are composed of both fine and coarse sand aggregates bound together with petroleum bitumen.

Most estate roads are laid with pre-cast concrete kerbs, fixed in advance of the pavement. British Standard 340: 1979 sets out standard sections and of these the most commonly used are: 5 × 10 in (127 × 254 mm) and the 6 × 12 in (152 × 305 mm), splayed or half battered. A 'half-section', half-battered kerb measuring 6 × 5 in (152 × 127 mm) is also included, for setting on top of the pavement.

Drawings of road edge details generally show a neat rectangular edge beam or foundation of concrete on which the kerb is bedded in mortar and backed up with concrete. A more practical detail is to set the kerb directly on top of a windrow or continuous mound of fresh concrete. This can be deposited directly into position from a truck mixer and will stay plastic long enough for the kerbs to be set accurately to line and level. Once the windrow concrete has set, the kerbs are backed up by a concrete haunch as normally (see Fig. 22.2). This alternative is much faster and obviates the need for forms in which to cast the foundation.

If the kerbs are to be fixed after the pavement is laid the full height of 254 or 305 mm is not required and, therefore, the half-section kerb is used. This is bedded on mortar into a groove or step

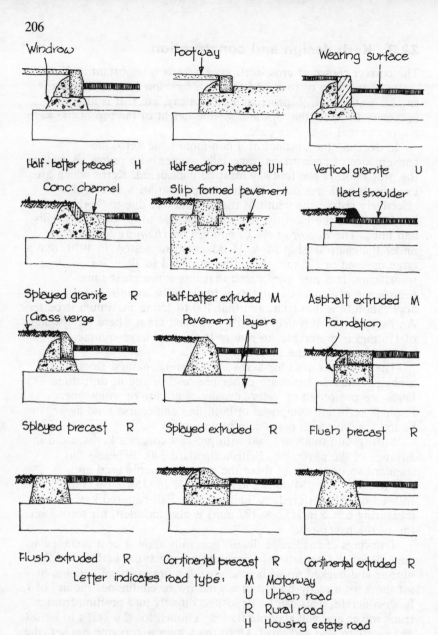

Fig. 22.2 Kerb constructions

in the formation and haunched with concrete as the full section (see Fig. 22.2). This construction makes it necessary to use formwork along the edge of the pavement and a special treatment to produce the step. It is not so satisfactory as the method using full depth kerbs, as these smaller kerbs can be more easily dislodged by frost or wheel contact. A bad practice, relying on pious hope rather than proven performance, is to lay a concrete pavement which is level right across, with a criss-cross of grooves scored down each edge, which are supposed to provide a key of sufficient strength to hold half-section kerbs bedded onto the slab in cement mortar. Invariably this key fails in many of the kerbs for one reason or another.

Figure 22.2 also shows a granite kerb set vertically and granite setts set to provide a splayed kerb. British Standard 435: 1975 covers the use of natural stones as kerbs. They are preferred for their greater durability, but laying is more difficult than with pre-cast concrete, because the stone is less precise in its shape and size.

Extruded kerbs, either of concrete or asphalt, are confined to major road works projects, where the cost of the equipment is commensurate with the size of the contract. Figure 22.2 shows a half-batter extruded concrete kerb which has been run into a groove formed in the pavement as it is laid. For these types of contract the pavement is usually laid as a continuous ribbon by a machine called a slip pavior.

Asphalt kerbs, also shown on Fig. 22.2, are only used along the back edges of hard shoulders and are extruded onto the finished surface which has been previously prepared by a tack coat of cationic emulsion.

Mountable or flush kerbs are splayed, flush or 'continental' kerbs and can be either pre-cast and set on an edge beam or windrow, or extruded in advance of the pavement. Figure 22.2 shows all these forms which, for the pre-cast kerbs, are similar in their construction to non-mountable kerbs but with the extruded kerbs they differ from their non-mountable counterpart in that they are laid on the sub-grade, not on the pavement.

22.4 Block paving

Where vehicle speeds are restricted to 50 km/h (about 30 mph) the use of concrete block paving as a road construction is finding favour. Typical applications are residential roads, building forecourts, car parks and drives, as well as precincts and landscaped areas.

The concrete blocks are pre-cast from high strength concrete, in a limited range of colours. They are accurately manufactured, either to a rectangular 100×200 mm module, or in an interlocking Z formation. They are laid dry on a carefully levelled bed of clean sharp sand, 50 mm thick, over a sub-base of soil-cement, cement-bound granular material, or lean-mix concrete. The blocks are tightly butted to prevent them from moving, and must be restrained at the edges of the pavement by an *in situ* or pre-cast kerb, or by a wall.

A concrete block pavement behaves like a flexible road of tar macadam, but has the durability of a concrete road plus qualities of colour and texture. The construction is dry, easily laid, with the minimum of plant and labour, it can be trafficked as soon as it is completed and, if necessary, areas can be taken up and relaid without marring the appearance.

Despite the fact that the blocks are laid dry on a sand bed, which would seem to afford natural drainage, it is necessary to lay block paving to the falls normal surface water drainage require because, in use, the joints between the blocks become sealed with fine dust and the surface is then virtually impervious.

22.5 Road drainage

There are two types of drainage associated with roads: sub-soil drainage and surface water drainage. The first is necessary where there is a high water table in the ground, which must be permanently lowered to maintain the stability of the sub-grade and pavement. The second deals with the disposal of rainwater run-off from the impervious carriageway.

The objective of sub-soil drainage is to reduce and hold the water table level to at least 1 m below formation level. This can be achieved either by land drains laid below the road, or, preferably, by French drains constructed along each side of the pavement (see Fig. 22.3). As shown in the illustrations, the French drains are constructed by laying pervious or open-jointed pipes at the bottom of a trench and then filling the trench with graded granular material and achieve a draw-down effect on the water table. They must be constructed to a greater depth than the required water table level to allow for the upward curvature of the draw-down line. The same consideration also applies to land drains, but since they are more closely spaced the height of the upward curvature is less. Where the road is in a cutting, the French drains would also be designed to receive any surface water run-off from the adjoining banks.

Surface water drainage, as explained in Ch. 21 of *Advanced Building Construction*, Vol. 1, consists of three parts: directing the water off the running surface to the kerbs, collecting of the kerb edge water into gullies and the disposal of the water in the gullies. The directing of the surface water is dealt with in section 21.6 of the above reference.

Once the surface water has reached the kerb channel, it is directed into a gully pot. This is of either stoneware or concrete, 300 to 600 mm in diameter, 600 to 1200 mm deep, bedded in and surrounded with concrete and finished on top with two courses of engineering bricks on which is set a cast iron frame and grate. Attached to the side of the gully pot is a trap with a rodding eye, the outlet of which is connected to a rainwater sewer (see Fig. 22.3).

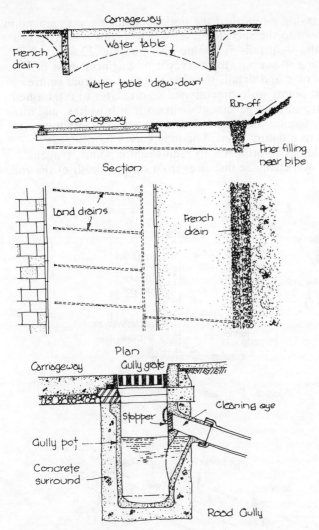

Fig. 22.3 Road drainage

22.6 Setting out roads and drains

While dissimilar in their nature and function, both roads and drains present the Surveyor or Site Engineer with similar problems. They both have to be set out to the correct line on the site and to the correct levels and gradients or falls. Additionally, the setting out of roads involves the construction of curves. In both cases the procedure usually adopted is to set out the centre line of the road or drain and from this set off-sets each side to give the road verge or drain trench. Having established the horizontal alignment, excavation can proceed down to the specified level which is constantly checked by a traveller and sight rails or boning rods (see Fig. 22.4).

Before starting to set out, the Surveyor or Engineer must be supplied with a plan and details showing the road or drain centre line intersection points with dimensioned co-ordinates to established reference points, all road pavement or drain trench widths, and all necessary levels (centre and edge of road pavement or drain invert levels), related to a datum point. For road curves he will also require dimensioned co-ordinates to radius centre points, tangent point locations and chainage distances from start to finish of the road section.

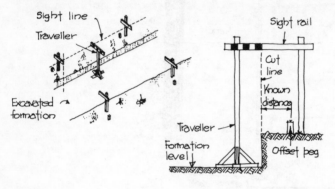

Road Profiles

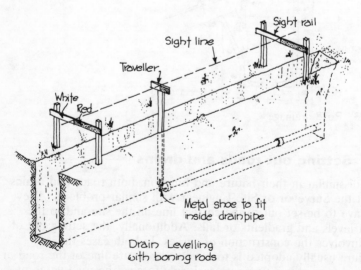

Drain Levelling
with boning rods

Fig. 22.4 Road and drain levelling

Having obtained all the information, the Engineer would then calculate all the co-ordinate angles and distances from the reference points to the centre lines that he is going to need and from this locate the centre line pegs. With a road, these pegs would be set at each end of each straight section, indicating the intersection point between the line of a straight and a curved section. In the case of a drain, these pegs would mark the centres of manholes. The usual practice with pegs is to drive in a 50 × 50 mm softwood peg at the position of the point and into the top of the peg drive a nail to give the precise location.

The next step is to set out and mark the required off-sets. In the case of a drain trench, the accuracy and permanence of this operation need not be of a high standard and the trench limits are often marked by a line of lime run along the ground just prior to the excavator starting work (when trench and excavator bucket are the same width only the centre line needs to be marked). Where roads are being set out, the offset points require to be more accurate and permanent and thus are marked by pegs which are each placed a uniform distance (usually 1 m) clear of the relevant cut line so as not to be disturbed by the work.

Setting out the offset pegs round the inside and outside of a curve is achieved by a process requiring a knowledge of surveying techniques and the use of a theodolite, but briefly it consists of calculating the length and angle of a series of small chords to the curve and locating the ends of these chords to give a series of points on the curve, known as chainage points. When the kerb is to be laid the Engineer would, by a similar process, establish the position of intermediate chainage pins between the chainage points and the kerb layer would stretch a string line between these points. Being on a curve, the back edge of the kerb would not follow this line and the kerb layer would be given the maximum deviation point which is at mid-chord length and is referred to as the mid-ordinate or versed sine (see Fig. 22.5).

As already mentioned, the levels of roads and drains are determined by sight rails and travellers. This method, however, can only be used for straight gradients and thus is always suitable for a drain, where any changes in gradient always occur at a manhole. In road construction the transition from one gradient to another is through a vertical curve. To set such a curve the Engineer would decide on the intervals he needs between chainage points – the sharper the curve the closer the points. He would then calculate the level each point would be if the gradient continued in a straight line, and from these he would deduct a calculated 'drop' distance to the vertical curve to give the reduced level – if the curve is a 'rise' (see Fig. 22.5). If the curve is a 'sag', i.e. the centre of the curve is above the line of the road, the calculated drop distances become rises and are added to the levels.

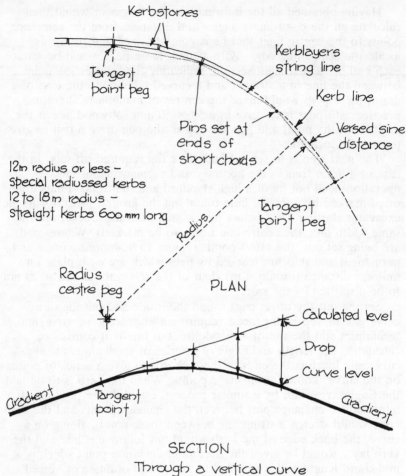

Fig. 22.5 Road curves

An alternative means of checking levels during excavation or pipe laying is by the use of a laser beam. This employs a visible light beam radiating from a laser head which can be made to rotate in a level plane or be fixed at a given line and angle. When set up and rotating, the laser head gives a visible plane at a fixed level which can be seen by the light striking a staff from which point level readings can be taken. Alternatively, if the head is fixed and tilted at a desired gradient, constant measurements taken downwards from the inclined plane would produce a uniform slope. For drains, the equipment can be set up on a manhole base at the drain gradient and the pipes levelled by adjusting their height until the beam hits the centre of a target disc standing within the pipe.

Chapter 23

Site protection and reinstatement

23.1 The need to protect

All sites, prior to a builder's arrival, are the product of nature, either alone or as controlled by the site owner. Even where it appears to be a wilderness, the area will contain natural features which, eventually, could enhance the appearance of the building. All too often these products and features are destroyed or disfigured by the process of erecting the building with which it was intended they should harmonise.

Trees, shrubs, plants and grass cannot be created even by our most advanced scientific knowledge and technology and therefore when constructing what we can achieve we should give due respect to what we cannot make, and protect these works of nature from harm.

Apart from these deferential considerations, the preservation of natural features which will make the finished work more attractive makes aesthetic and economic sense. It takes a long time to re-establish trees, shrubs and grass, and thus complete the original overall design. If, by careful protection the need to buy and plant these works of nature is removed, the cost of the builder's work is correspondingly reduced.

23.2 Methods of protection

The best way to protect natural features on a site is to prevent any person or any machine going near to them. If, by sturdy fencing, the areas of the site which are to be retained as they were can be completely separated from the areas where building operations are in progress, the problem is solved. Unfortunately this cannot be done on many sites because the area required for the building work is so much greater than the eventual site coverage of the structure. Where complete segregation cannot by achieved, the major features can sometimes be provided with their own fences to keep them safe, but when lack of space prevents even this, then they should be enclosed by a protective timber structure at least 2 m high to prevent accidental damage.

Trees present a problem where space is restricted, not so much from the protection of the trunk from collision damage, but the avoidance of damage to the branches by large vehicles or equipment. Where the risk of this is very high, it is better to cut and seal the threatened branch than to allow it to be broken, thereby inviting diseases or decay to start. In really difficult cases it is possible to dig up the tree, retaining an adequate root ball, and transport it to a new position, either permanently or temporarily. The successful execution of this type of operation requires the skill and specialised equipment of experts in the subject.

Grass presents less of a problem since it is easier to reinstate than a mature tree but, nonetheless, if the disturbance is only to be for a few weeks, then it is worth considering carefully stripping the turf (if it is of a good quality), folding the turves inwards and stacking them for future replacement. An occasional watering of the stack in dry weather will prevent the grass dying off. It is only possible to stack turves successfully for a period which is much shorter than the length of most building contracts and where the grassed area is to be used for other purposes, say the materials compound, for the duration of the contract, it must be accepted that new grass will have to be provided at completion. In anticipation of this it is a wise precaution to strip the topsoil into a convenient heap, so that this can be replaced before the grass seed is sown or the turf laid, thereby providing a growing medium uncontaminated by building materials or rubbish.

23.3 Tree planting

In no case should any work be undertaken if the ground is very wet or frozen, although the best time to transplant trees and shrubs is in the winter, when they are dormant.

The ground should be prepared by digging it thoroughly for a good distance round the planting position and very heavy or very light soils should be treated with well-rotted manure or garden compost. A planting hole must be dug of sufficient depth and width to accommodate the roots when fully spread out, the tree carefully placed in it and good quality soil returned and firmed round the roots. In most cases, the newly planted tree will require a stake to hold it until the root system has become established. This stake should be driven in before the tree is planted, to avoid damage to the roots, and there should be a pad between the stake and the tied stem to prevent chafing of the bark on the stake.

If properly planted and supported, trees should need no more after-care than watering, if dry weather follows their being planted, and an occasional inspection of the ties to check they are still secure, or to ease them as the tree grows.

23.4 Grass planting

There are two ways to obtain a lawn or grassed area: by sowing grass seed or by laying turf. In either case the first operation is to clear away all rubbish and building materials and to rake over the area to remove all stones, broken brick and tile and any other small objects over 30 to 40 mm in diameter. The topsoil should then be prepared by Rotavating, levelling, rolling and raking. Alternatively, if the topsoil was removed, the ground surface should be cleared of rubbish and broken up to ensure free drainage, and the topsoil returned, levelled and raked. Whether grass is to be sown or turf laid it is advisable to apply an appropriate fertiliser first.

Seed should be sown between mid-March and May or mid-August to October at the rate of 50 g/m^2 and the surface lightly raked. There are eight different varieties of grass used in lawns, mixed in varying proportions according to the type of use the area will receive. Lawn seed mixtures for general purposes contain 50 per cent of ryegrass to make it a hardwearing lawn, but this is a coarse grass and is entirely omitted from seed intended for fine lawns.

Turves are laid between the end of September and early March. They are cut into 300 × 900 mm rectangles and are laid in rows with the joints staggered in each row, each pressed or lightly beaten into place. When the laying is complete, the area is dressed with fine soil brushed into all the crevices.

On steep banks both the topsoil in which the seed is sown or the turves if used will require holding in position until a root growth has been established. A simple way to do this is to peg down a nylon or similar net, which will prevent any movement and through which the grass can grow. The net is left permanently in position and can be mown over with safety. Turf, which should be laid diagonally on a bank, can be held by a net or each turf can be retained by a wooden peg driven into the middle to within 25 mm of the surface. These pegs should be removed once the root growth is firm and before the bank is mowed.

23.5 Maintenance

In most contracts it is the Contractor's responsibility to ensure that
newly planted trees are not allowed to die or be damaged and that
grass is cut and watered until the handover date. If, during the
defects liability period, a tree dies or becomes diseased, it is
generally included in the Schedule of Defects at the end of the
period and the Contractor is then faced with the cost of replacing it
with another.

Bibliography

National Federation of Building Trades Employers *Construction Safety*. B.A.S. Management Services.

The Federation of Civil Engineering Contractors, *Guide to the Construction Regulations 1961 and 1966*. D. Sharp & Co. (Printers) Ltd.

M. J. Tomlinson, *Foundation Design & Construction*. Halstead Press.

Mitchell's Building Construction Series. Batford Ltd.

R. Chudley, *Construction Technology* Vol. 4. Longman.

R. P. Johnson, *Composite Structures of Steel and Concrete*. Granada.

H. Werner Rosenthal, *Structure*. Macmillan.

Fred Angerer, *Surface Structures in Building*. Alec Tiranti.

A. M. Haas, *Thin Concrete Shells*. Wiley.

American Institute of Timber Construction *Timber Construction Manual*. Wiley.

Walter L. Slater, *Floors and Floor Maintenance*. Applied Science Publishers.

Arthur W. Birley and Martyn J. Scott, *Plastics Materials*. Longman.

R. W. Murphy, *Site Engineering*. Longman.

Transport and Read Research Laboratory, *The Design and Performance of Road Pavements*. HMSO

Bibliography

Index